Mass Spectrometry
A Foundation Course

Mass Spectrometry
A Foundation Course

K. Downard
University of Sydney, Sydney, Australia

RS•C

advancing the chemical sciences

ISBN 0–85404–609–7

A catalogue record for this book is available from the British Library

Published by The Royal Society of Chemistry
Thomas Graham House, Science Park, Milton Road,
Cambridge CB4 0WF, UK

Registered Charity Number 207890

For further information see our web site at www.rsc.org

Typeset by RefineCatch Ltd, Bungay, Suffolk, UK
Printed by TJ International Ltd, Padstow, Cornwall, UK

To Craig

"The occasional fineness of line, the masterly distribution of masses"

(from The Master (1895) by Israel Zangwill (1864–1926))

Preface

This book presents a broad coverage of the theory and application of mass spectrometry to provide the reader with an appreciation and understanding of the importance of mass spectrometry across a range of scientific disciplines. It is uniquely organised to enable a course or unit in mass spectrometry to be constructed at either the undergraduate or postgraduate level for students of a range of backgrounds and educational experiences where no single course can be deemed suitable for students of the physical, chemical, environmental, biological and medical sciences.

It is published at a time when most available textbooks present an introduction to mass spectrometry, a broader treatise devoid of much detail or one that is focused on a particular area of the field. A large number of multiple author collections describing specialised disciplines, often inspired by a conference or workshop, together with new encyclopedic series have provided readers with up-to-date descriptions of mass spectrometry research and applications though usually in a less cohesive and accessible format. This has left a new scholar with some difficulty in comprehending the foundations, role and capabilities of mass spectrometry.

This has motivated the construction of a new book on mass spectrometry that presents a broad treatise of the field across a wide range of scientific disciplines in a single accessible and affordable volume. Sufficient depth is presented throughout the book to enable students to understand the principles behind and the reasons for particular experiments, together with ample representations of mass spectral data and applications. Importantly, the book provides a reference text around which a series of university level courses can be constructed for the education of students with varied backgrounds, experiences and interests.

The unique design of the book achieves this through the presentation of core sections that are common to all mass spectrometry experiments. These sections are coupled to content from other optional sections and specialised chapters dependent upon a student's educational level, specialisation and interests. Recommended course structures are presented in the front of the book. At the same time, the organisation of the book is designed to present the field of mass spectrometry in a logical manner regardless of the course undertaken. Specialised chapters are included on organic mass spectrometry, ion chemistry, biological mass spectrometry featuring proteomics, mass spectrometry in medical research, the environmental and surface sciences and accelerator mass spectrometry.

Large numbers of mathematical equations and derivations have been avoided and the theoretical description of mass spectrometry based experiments has been kept to a minimum. The absence of a large number of citations to the enormous body of published research on mass spectrometry was also deliberate, not so as to ignore the important work contributed by many scientists throughout the world, but rather to prevent the reader from being distracted by extensive annotations and references throughout the course of the text. Each chapter concludes with a list of key references and recommending reading material providing a springboard to further study.

The author hopes that this book will assist with the teaching of mass spectrometry to the field's future pioneers. Certainly, mass spectrometry education will remain an important exercise given the important, and in some cases essential, role that mass spectrometry plays in scientific discovery.

Kevin M. Downard

Contents

Guide to a Foundation Course in Mass Spectrometry

	Chemistry	Physics	Biology	Medicine	Environmental Sciences
Undergraduate Core	1.2.2 1.3 2.1 2.2.1 2.3–2.5 3.1 3.2.1–3.2.3 3.2.5 3.26 3.2.8 – 3.2.10 3.3 3.3.1 3.3.2 4.1 5.3 5.4	Ch 1 2.1 3.1 3.2.1 4.1 6.2 6.3 9.4–9.6 Ch 10	2.1 2.3–2.5 3.1 3.2.6–3.2.9 3.2.10 3.2.11 4.1 7.1 7.2.1–7.2.3	2.1 2.3–2.5 3.1 3.2.1–3.2.3 3.2.6 3.2.8 3.2.9 Ch 8	2.1 2.2.1 2.3–2.5 3.1 3.2.1–3.2.3 3.2.5 9.1–9.3 9.6 10.1 10.3 10.4
Undergraduate Optional*	1.1 1.2.1 3.3.3 4.2.1 4.2.2 5.1 5.2 Ch 6	2.2–2.5 3.3–3.6 4.2 6.1	3.2.4 3.3 4.2 7.2.4–7.2.6 7.3 7.4	3.2.7 3.2.10 4.1 5.3 7.1 7.2.1–7.2.3 10.1 10.3 10.5	4.1 5.3 10.2
Graduate/ postgraduate Supplement**	2.2.2–2.2.5 3.4–3.6 4.2–4.7	3.2 4.3–4.7 6.2 6.3	2.2 Ch 8	2.2 3.2.11 4.2.1–4.2.3 7.2.4–7.2.6 7.3–7.4	2.2 9.4 9.5

All sections listed represent the entire section (with sub-sections). All subsections listed represent the entire subsection only.
* optional sections and subsections should be added to the core material in the order that they appear in the text, not the order they appear in this table. ** postgraduate material should be taught in addition to, or as a supplement for, the undergraduate material dependent on the exposure of students to this subject matter at the undergraduate level.

Acknowledgements

I owe a particular gratitude to John Bowie for introducing me to the exciting field of mass spectrometry and for his support throughout my career. John's internationally recognised research in gas phase ion chemistry instilled in me an early appreciation of the positive aspects of negative ion mass spectrometry beyond the analytical.

I am grateful to many colleagues and students past and present, too numerous to mention here, in both the mass spectrometry community and further afield who have contributed to my own education and challenged my teaching of mass spectrometry. I also thank my wife and family for their love and support.

Finally, it has been my pleasure to work with Janet Freshwater, Robert Eagling, Tim Fishlock and the entire editorial and publication team at the Royal Society of Chemistry. My thanks also go to Edward Abel, former president of the society, who catalysed this interaction. Their belief in, and support of, this project has made the book possible.

Kevin M. Downard

Mass Spectrometry's Beginnings

1.1 A BRIEF HISTORY

1.1.1 Early Pioneers and Cathode Rays

Mass spectrometry had its beginnings in experiments performed over a century ago. Scientists in the late 19th century began conducting experiments within evacuated glass tubes in order to gain some understanding of the nature of electricity.

George Johnstone Stoney was the first to report that electricity has its basis in a particle, or an "atom of electricity" that he referred to as an electron. Stoney measured the charge of the electron in 1894 but it was left to Joseph John (J.J.) Thomson to measure the charge-to-mass ratio (*e/m*) of the electron and estimate its mass at a thousand times less than that of a hydrogen atom. Thomson had developed an interest in the electron while investigating the passage of electricity through gases in his laboratory in Cambridge. Thomson believed that the stream of rays emitted from a negatively-charged cathode, known as *cathode rays*, consisted of these particles. He also proposed that the particles (which Thomson preferred to call *corpuscles*) were one of the bricks from which all atoms were built – a controversial theory at the time. Thomson went on to describe his case in the book *Corpuscular Theory of Matter* published in 1907.

1.1.2 Positive Rays

Some time earlier Eugen Goldstein, a scientist in Germany who had given the name to cathode rays and studied them for several decades, discovered that the presence of gases in *cathode ray tubes* also gave rise to rays that behaved very differently from cathode rays (Figure 1.1). Wilhelm Wein in 1898 was able to deflect these rays in the opposite direction to cathode rays using magnetic and electrical fields. He

1

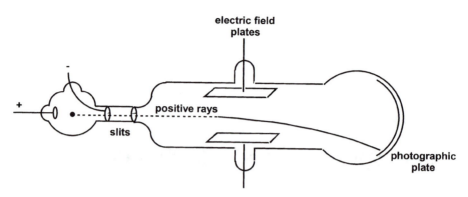

Figure 1.1 *Representation of the cathode ray tube showing the deflection of positive rays (ions)*

concluded that they were the positive equivalent of the negatively-charged cathode rays and carried "positive electricity". Wein's experiments, however, showed that these rays contained particles with masses much larger than an electron and of the order of a hydrogen atom. Fascinated by this positive form of electricity, Thomson improved upon the data of Wein by operating his cathode ray tube at lower pressures and showed also these positive rays could be deflected from a straight line by perpendicular electric fields onto a photographic plate (Figure 1.2).

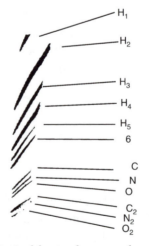

Figure 1.2 *Parabolic paths of ionised forms of atoms and small molecules within a cathode ray tube*
(Source: Fig. 5, plate 1, J.J. Thomson, *Recollections and Reflections*, G. Bell and Sons Ltd., London, 1936)

1.1.3 The First Mass Spectra

First with hydrogen, and later with other atoms and molecules of carbon, nitrogen and oxygen, Thomson discovered that each charged particle followed its own parabolic path based upon their detection on the photographic plate. He reasoned that no two particles would strike the plate at the same place unless they possessed the same velocity and charge-to-mass ratio. Thomson quickly realised that an inspection of the plate showed at a glance how many kinds of particles there were in the rays and that, by knowing the value of *e/m* for one parabola, the values of *e/m* for all the others could be deduced.

In his recollections, published in 1936, four years before his death, Thomson recounted that the positive rays contained atoms and molecules of all gases, common elements or molecules present, and correctly suggested that the positively charged particles were formed by the loss of an electron. In other words, the positive rays consisted of charged atoms or molecules, known as *ions*. Thomson concluded that the positive ray spectra possessed many advantages over other approaches for chemical analysis. The number of components as well as their atomic or molecular weight could be measured from these spectra. He argued that by using long exposures the approach could be "exceedingly delicate" allowing for the presence of a trace of gas to be detected "too small in amount to be measured by any other spectroscopic method". Another noted advantage was that the method was not dependent on the purity of gas analysed. Impurities merely appeared as additional parabolas in the spectrum and did not contribute errors to measurements of atomic or molecular weight.

Thus the field of mass spectrometry was born and these important features, recognised by Thomson, remain to this day. Mass spectrometers are able to:

(i)   measure the atomic and molecular weights of charged species in complex sample mixtures,
(ii) analyse compounds at extremely low sample levels,
(iii) analyse compounds in a mixture without purification of that mixture.

These advantages, and many others, considerably outweigh the few disadvantages described later such as the loss of the sample once it is analysed. But first the work of Thomson's student Francis Aston is briefly reviewed. Aston's work led to the discovery of isotopes which have important implications for mass analysis.

1.2 ISOTOPES AND THEIR IMPLICATIONS FOR MASS MEASUREMENT

1.2.1 Discovery of Isotopes

Commencing in 1909, Francis Aston accepted an invitation to work as an assistant to Thomson in Cambridge to study positive rays. It was during this period that improvements to Thomson's cathode ray experiments were made and several new "mass spectrographs" were constructed. These instruments, forerunners of the modern mass spectrometer, enabled Aston to separate two isotopes of elemental neon, ^{20}Ne and ^{22}Ne. The principle was extended to other chemical elements and led to the discovery of 212 naturally occurring isotopes. From this work, Aston formulated the so-called "Whole Number Rule" that states when the mass of the oxygen isotope is defined, all other isotopes have masses that are nearly whole numbers or integers. Carbon, rather than oxygen, is now considered the standard isotope mass upon which all other isotopes are measured.

1.2.2 Isotopes and Mass Measurement

Isotopes are atoms of the same element with different numbers of neutrons in the atomic nucleus. These add mass but not charge to the atoms and molecules composed of them. Naturally occurring carbon, for example, is a mixture of two isotopes ^{12}C and ^{13}C that represent 98.9% and 1.1% of all the carbon on Earth. The individual carbon atoms have masses of either 12.00000 (assigned as a standard by the International Union of Pure and Applied Chemistry, IUPAC) or 13.003354 mass units (u, formerly atomic mass units *amu*). The average mass of carbon is calculated as 12.011 u based on these masses and the natural occurrence of each isotope ($98.9/100 \times 12.000 + 1.1/100 \times 13.003 = 12.011$). Isotope masses and relative abundances for all the common elements are provided in Appendix 2.

If a mass measurement is made where the isotopes of each atom are separated, the mass will reflect a *monoisotopic* (or one isotope) *mass*. If the isotopes are unresolved, the mass will reflect an *average mass*. Note that the average mass value is larger than the monoisotopic mass for the lightest isotope since it contains a contribution from the heavier isotopes.

Unlike Thomson's early mass spectra, modern mass spectrometers record both the mass-to-charge ratios of ionised forms of atoms or molecules and also their *relative abundance* in the spectrum. Thus when the isotopes of an atom are separated or resolved in a mass spectrum, the relative intensities of the ions reflect their relative levels. To see this, consider the mass spectrum of an atom of chlorine. Chlorine has two isotopic forms ^{35}Cl and ^{37}Cl. The mass spectrum for an ionised chlorine atom will therefore appear as shown in Figure 1.3 in a bar graph representation.

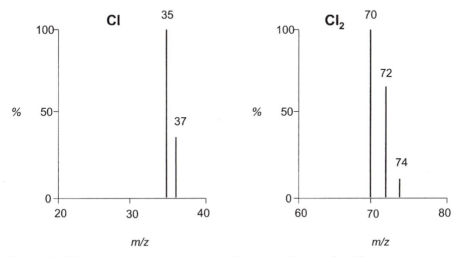

Figure 1.3 *Electron ionisation mass spectra of atomic and molecular chlorine*

Note that the relative height or abundance of the ^{37}Cl isotope is one third of that of ^{35}Cl. This is consistent with the relative abundances of the two isotopes which are 75.78% and 24.22% for ^{35}Cl and ^{37}Cl respectively (Appendix 2).

Consider a molecule of chlorine gas Cl_2. A molecule can be formed either from two ^{35}Cl atoms, one ^{35}Cl and one ^{37}Cl atom, or two ^{37}Cl atoms. The mass spectrum of the ionised form of the molecule reflects this and shows three resolved ionic forms (at *m/z* 70, 72 and 74) with relative abundances of approximately 100, 64 and 10% respectively (Figure 1.2). These values can be derived theoretically by multiplying the relative abundance for each atom and normalising the values to 100. The relative abundance for chlorine molecules comprised of mixed isotopes must be multiplied by two in order to account for the two forms possible; that is $^{35}Cl^{37}Cl$ and $^{37}Cl^{35}Cl$.

It is a simple matter to calculate the relative height of isotope peaks for any particular molecule. In the case of an isotope of one integer unit of mass higher than the molecule (*i.e.* M + 1), the relative height of this peak can be expressed as a percentage given by equation 1.1. In this equation, *%I* represents the percentage of naturally occurring isotope one mass unit greater than the common isotope for each element, E. The number of atoms of that element in the molecule is defined by n.

$$\text{ratio of (M + 1)/M peak} = \Sigma_E (\%I \times nE)\ \% \qquad (1.1)$$

Hence for a molecule of ethanol (C_2H_5O), the ratio of the (M + 1)/M isotope peaks will be equal to $(1.08 \times 2) + (0.015 \times 5) + (0.038 \times 1)$ or 2.27% based upon the percentages of naturally-occurring ^{13}C, ^{2}H and ^{17}O (see Appendix 2).

In general, the contribution of any element with two common isotopes (of a and b percent) to an isotope peak (M + N)/M is given by equation 1.2.

$$\text{ratio of (M + N)/M peak} = N!(a)^{n-1}(b)^n/N(a)^n \qquad (1.2)$$

Hence the relative peak height of (M+2)/M for a molecule of Cl_2 is calculated to be $2(0.76)^1(0.24)^2/2(0.76)^2 = 0.076$. Note that N! (or N factorial) equals the multiple $N \times (N-1) \times (N-2) \times \ldots \text{to} \times 1$.

1.3 MOLECULAR WEIGHT

The molecular weight of a compound is the sum of the atomic masses for all atoms in the molecule weighted according to the relative abundances of their isotopes. For example, the molecular weight of a molecule of sucrose or common sugar ($C_{12}H_{22}O_{11}$) is the sum of the atomic masses for 12 carbon atoms, 22 hydrogen atoms and 11 oxygen atoms. The average molecular weight calculated from the atomic weights for these atoms is 342.299. If the isotope peaks for the compound were resolved or separated in a mass spectrum, the expected mass of the monoisotopic ions or ions containing only the lightest isotopes of each element would be 342.116.

Measuring molecular weight is a fundamental activity of modern mass spectrometry, but as we shall see later, mass spectrometers are used for many other purposes including the complete structural characterisation of molecules, the quantitation of components within complex chemical and biological mixtures, fundamental investigations of ion behaviour and reactivity studies of molecular complexes, and even the radiodating of archaeological relics.

1.3.1 Elemental Composition and Mass Accuracy

The manner in which accurate mass measurements are obtained is described later in Chapter 5 in the context of the instrumentation required, but assume for the moment that the molecular weight of a compound can be measured with a high degree of accuracy. Subject to the accuracy obtained, it may be possible to assign the elemental composition to a compound based solely on the molecular weight measurement. This is because the monoisotopic atomic masses for the elements are not exact integers; rather they have a fractional mass component. Consider for example a small organic molecule whose molecular weight based on a measure of the mass-to-charge ratio of its monoisotopic ions is 46.0419. This molecular weight value is consistent with a molecule of ethanol with an elemental formula of C_2H_6O, but not for instance a molecule of nitrous oxide NO_2 with a molecular weight of 45.9929. The mass accuracy required to distinguish these two molecules can be calculated as one-half of the difference between their molecular weights divided by the molecular weight of one of them. In this example, a mass accuracy of 0.00053 or 530 parts-per-million (ppm) is required. The division of the difference by two arises since if either molecular weight is in error by greater than one-half of the difference, the measurement will not enable the two elemental compositions to be distinguished.

Such determinations are not limited to small molecular weight compounds provided that sufficient mass accuracies can be obtained. It has recently been shown using an ion cyclotron resonance mass spectrometer that two peptides with molecular weights that differ by just 0.00045 Da could be distinguished. This mass difference corresponds to a value less than the mass of a single electron (Figure 1.4).

1.3.2 Nitrogen Rule

The *nitrogen rule* is of use in assigning an elemental composition to a compound based upon its molecular weight. For any compound that contains only C, H, N, O, S, Si or halogen atoms (F, Cl, Br or I), the nitrogen rule states that the *nominal molecular weight* for the compound will be even only if the number of nitrogen atoms is an even number or zero. The nominal molecular weight represents only the integer portion of the value. In the case of ethanol, the nominal molecular weight is 46. This is an even number consistent with the lack of nitrogen atoms in the molecule. The value is inconsistent with the one nitrogen atom present in NO_2. Thus based on the nitrogen rule, we can assign the correct elemental composition of C_2H_6O even where a mass accuracy of less than 530 ppm is unattained. A compound containing an odd number of nitrogen atoms in addition to atoms of any of those elements detailed above, will have a odd nominal molecular weight. In the case of NO_2, the nominal molecular weight is seen to be 45.

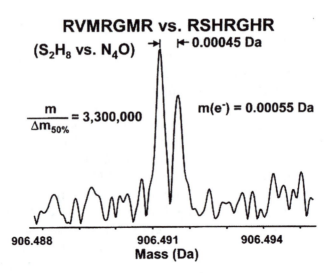

Figure 1.4 *ESI FT-ICR mass spectrum of [M+2H]²⁺ ions of two peptides that differ in mass by 0.00045 Da. The mass resolution at FWHM is calculated to be 3.3 million*
(Source: Fig. 2, Fei He, Christopher L. Hendrickson and Alan G. Marshall, *Anal. Chem.*, 2001, **73** (3), 647–650)

1.3.3 Double-Bond Equivalents

An additional rule of use in assigning a correct elemental composition is the *double-bond equivalents* (DBE) or number of "double bonds plus rings" rule. Put simply, this rule calculates the number of double bonds or aromatic rings in an organic molecule with the elemental composition $C_xH_yN_zO_n$. The number of DBE in such a molecule is defined by equation 1.3.

$$DBE = x - y/2 + z/2 + 1 \qquad (1.3)$$

In the case of ethanol described above, the DBE calculated based on the elemental composition C_2H_6O equals zero, a value consistent with only single bonds being present in the molecule. For a molecule of benzene with an elemental composition of C_6H_6 the DBE equals 4 [or $6-(6/2) + 1$]. This is confirmed by its structure since three double bonds are associated with the aromatic ring and a further DBE is associated with the ring itself.

FURTHER READING

J.J. Thomson, *Recollections and Reflections*, G. Bell and Sons Ltd., London, 1936.

F.W. Aston, *Mass Spectra and Isotopes*, Edward Arnold and Co., London, 1933.

J.L. Putman, *Isotopes*, Penguin Books, 1960.

R.B. Firestone, S.Y.F. Chu and C.M. Baglin, *Table of Isotopes*, 8th edition, John Wiley and Sons, New York, 1998.

CHAPTER 2

The Mass Spectrum

2.1 CONCEPT OF CHARGE AND THE MOLECULAR ION

To this point, only the molecular weight of a molecule has been considered to contribute to its appearance in a mass spectrum. However, as recognized by Thomson, the particles detected in mass spectrometry are ions. Thus they have both mass and charge, the latter of which is important to their detection. Mass spectrometers are unable to detect neutral molecules and radicals; a charge must be imparted onto an atom or molecule before it can be studied. The reason for this is that charged molecules can be "handled" or their paths controlled through the use of electric and magnetic fields, while radicals and neutral molecules are unresponsive.

Depending on the nature of the ionisation process, and also the nature of the atoms and molecules themselves, different ion types can be formed. The most common and traditional way in which ions are produced in a mass spectrometer is through the loss of an electron (equation 2.1). This often occurs by the initial collision of a gaseous atom or molecule with an electron in a process known as *electron impact* or *electron ionisation* (EI) (Chapter 3).

$$M + e^- \rightarrow M^{+\cdot} + 2e^- \tag{2.1}$$

The product ion formed is known as a *radical cation* as it is an odd electron species with a positive charge. Since the product, $M^{+\cdot}$, has the same mass as the molecular weight of the compound M from which it was produced, it is known as a *molecular ion*. In general, the notation $^{+\cdot}$ adjacent to a molecule's structure indicates that the molecule is deficient of an electron without designating the site of the charge.

The charge of the ion is equal to the charge of an electron (e) defined as 1.6×10^{-19} Coulombs (C). Were the ion to possess two charges, perhaps through loss of electrons from two atoms or groups of atoms within the same molecule, the charge on the ion would be $2e$. Thus the charge of an ion is always some multiple of the charge of an electron or ze, often just denoted z. In a mass spectrum, ions appear at a mass-to-charge ratio defined m/z, where m is the mass of the ion and z is the charge. Since z often (though not always) has a value of one, early references in mass spectrometry refer to an ion's mass-to-charge ratio as m/e.

It is important to note that not all ions are formed by electron loss. Some electronegative atoms or molecules can attract an electron during electron impact (equation 2.2). These ions, denoted $M^{-\bullet}$, are still referred to as molecular ions. The same is true of even electron species formed by the adduction or loss of a charge carrying atom or group. An ion may be formed for instance by the protonation of a molecule. $[M + H]^+$ is also referred to as a molecular ion, or more strictly a *quasi* or *pseudo-molecular ion* since the mass of the ion is now larger than the molecular weight of the molecule by one atom of hydrogen.

$$M + e^- \rightarrow M^{-\bullet} \qquad (2.2)$$

In general terms, most ions formed, dissociated and studied in mass spectrometry are positively-charged ones. This is because their production is usually more efficient for most classes of compounds over their negatively-charged counterparts. This is not to say that negative ion mass spectrometry is not important or not used; indeed many important observations and applications are based on these experiments (see for example Chapter 7). The polarity of the lenses used to repel ions from the source is the same as that of the ions themselves. The polarity of the voltage applied to the detector, on the other hand, is opposite to that of the ions in order to attract them to the ion detector. This leads to the detection of either positive or negative ions, but not both simultaneously.

2.2 FRAGMENT IONS

2.2.1 Formation of Fragment Ions

If sufficient energy is deposited into the molecule during ionisation, the molecular ions may dissociate into smaller mass fragments that themselves may be ions. For example, an odd-electron radical cation, $M^{+\bullet}$, may dissociate to form two fragments one of which is an even electron fragment of product ion, and the other is a radical $R^\bullet$ (equation 2.3).

$$M^{+\bullet} \rightarrow F^+ + R^\bullet \qquad (2.3)$$

Alternatively, $M^{+\bullet}$ could dissociate to produce a smaller mass fragment and a neutral molecule N (equation 2.4).

$$M^{+\bullet} \rightarrow F^{+\bullet} + N \qquad\qquad (2.4)$$

Only the ionic products F^+ and $F^{+\bullet}$ are passed through the instrument and detected. Beyond simple bond cleavages, fragment ions can be produced following the rearrangement of atoms if sufficient energy is available to facilitate bond cleavage and formation. Hydrogen atoms and protons, for instance, are frequently transferred from a remote site to the ionic centre prior to cleavage of the molecular ion.

The most abundant ion peak in a mass spectrum, whether it is that of the molecular ion or that of a fragment, is referred to as the *base peak*. Ion abundances are measured relative to the intensity of the base peak, set arbitrarily to 100%, and usually plotted to the top of the *y*-axis (Figure 2.1).

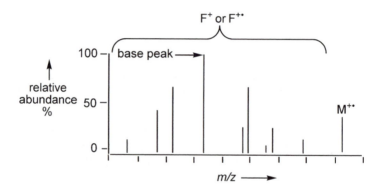

Figure 2.1 *Representation of an EI mass spectrum for molecule M*

2.2.2 Stability of Fragment Ions

The relative stability of the fragment or product ion is a major factor that influences the appearance of fragment ions in a mass spectrum. For a fragment ion to appear in a mass spectrum, it must be produced from a significant proportion of the precursor ions and be relatively stable to further fragmentation. The most predominant fragmentation pathways are those with the lowest energy barriers to product and with the lowest change in free energy (ΔG, where ΔG is associated with changes in enthalpy, ΔH, and entropy, ΔS, according to $\Delta G = \Delta H - T\Delta S$). Other factors, including the strengths of the bonds broken during the fragmentation process and the time allowed for dissociation, also influence the appearance of fragment ions in the mass spectrum.

A few factors that can stabilise a fragment ion are now considered. Although illustrated for simple organic ions, these effects can be extended to other classes of compound.

2.2.3 Stabilising Effects

One of the most common ways that positive ions can be stabilised is through electron donation from a neighbouring atom or group of atoms. The transfer of electron density toward a positive centre is known as the *inductive effect*. Alkyl groups in hydrocarbons have the ability to donate electrons to a electron-deficient positive ion centre, such that the stability of carbocations follows the order: $(CH_3)_3C^+ > (CH_3)_2CH^+ > CH_3CH_2^+ >> CH_3^+$. Thus the fragmentation of hydrocarbons gives rise to mass spectra in which ions at m/z 57 $((CH_3)_3C^+)$ and 43 $((CH_3)_2CH^+)$ are more abundant than those at m/z 29 $(CH_3CH_2^+)$ and 15 (CH_3^+). In contrast, electron-withdrawing atoms (F, Cl, Br, I) or groups (–OH, –NO$_2$) have a destabilising effect on a neighbouring positive ion centre.

A second stabilising effect is the *mesomeric effect*. Here a positive ion centre is stabilised by its conjugation with multiple (unsaturated) bonds. Hence the ion is stabilised through *delocalisation* of the charge across the molecule or fragment. As an example, the ion $CH_2 = CH–CH_2^+$ (m/z 31) can be stabilised by charge delocalisation to the form $^+CH_2–CH = CH_2$. This delocalisation of charge through *bond resonance* also stabilises a phenyl ion (equation 2.5). Ions at m/z 77 $(C_6H_5^+)$ are a signature of aromatic compounds in EI mass spectra.

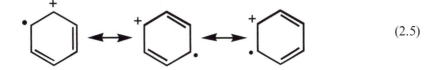

$$(2.5)$$

2.2.4 Quasi-Equilibrium Theory

The unimolecular decomposition of molecular ions into fragments can be explained by the Quasi-Equilibrium Theory (QET). The QET provides a theoretical description of how these processes take place inside a mass spectrometer. The ionisation of a molecule by electron loss (equation 2.1) or electron capture (equation 2.2) occurs within approximately 10^{-15} s, a time much shorter than that required for a molecular vibration. Hence the ionisation event can be considered to be a "vertical transition" with no change in internuclear distances (Figure 2.2). If the geometry of molecular ion differs from that of the neutral molecule,

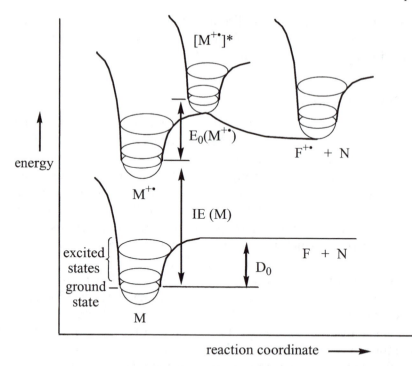

Figure 2.2   *Schematic representation of energy levels associated with the ionisation and dissociation of a polyatomic molecule M along a reaction coordinate*

the states for the latter will appear slightly along the reaction co-ordinate (to products) and the transition is said to be non-vertical or *adiabatic*.

Since pressures within the ion source are typically very low, each ion can be considered to be an isolated system as ions do not collide with each other or background gas molecules. If the excess energy transferred to a molecule during the ionisation process is distributed between all possible excited states in the ion, and these excited states interconvert between one another, a *quasi-equilibrium* is said to exist among these states. Therefore it is postulated that each excited state is of equal probability and that an ion's dissociation pathway depends only on its internal energy and the structure of the molecule, and not upon the initial site of ionisation or the nature of the ionisation process.

Few exceptions have been found to the QET and such dissociations are termed *non-ergodic*.

The *ionisation energy* (IE) is defined as the minimum energy necessary to produce a molecular ion, $M^{+\bullet}$ from its ground neutral state, M. It can be determined by raising the energy of the electrons in an electron impact source until a molecular ion is detected. Since the energy is related directly to the potential through which the electrons are accelerated, it is often referred to as the *ionisation potential*.

The appearance of a fragment ion in a mass spectrum occurs if there is sufficient energy to ionise the neutral molecule and overcome the activation barrier. This is referred to as the *appearance energy* (E) of the fragment ion. An additional factor, the *kinetic shift*, may contribute to the appearance of a fragment ion and results from the excess energy required in order for the fragmentation process to occur within the time the molecular ion spends in the ion source. The kinetic shift increases with the size of the molecule, the activation energy and the "tightness" of the transition state.

Where there is no activation barrier and the kinetic shift is zero or negligible, the minimum energy required for (M) to dissociate into fragments can be defined as IE. All molecular ions with internal energies less than E_0 do not dissociate regardless of the time available for fragmentation. The appearance energy of a fragment can be defined by equation 2.6.

$$E = IE(M) + E_0 \qquad (2.6)$$

In the case in which two dissociation pathways are possible where $(E_0)_1 \approx (E_0)_2$, the pathway which proceeds through a "loose" transition state, $[M^+]^*$, will predominate. This is because the energies of the excited states of a transition state in which the components are loosely associated will be similar to one another, over those for a "tightly-activated" transition state, such that they are easier to surmount. Ion dissociations involving the breaking of a single bond (with a bond dissociation energy of D_0) usually proceed *via* a loose transition state complex and are typically favoured over rearrangement reactions that involve tighter transition states. Computer algorithms are available to calculate the energies of such states for low values of $(E - E_0)$. Difficulties, however, arise for a real molecular system in predicting an ion's physical parameters, including its energy for activation.

Both theoretical and experimental results have shown that the rate constant k for a fragmentation process increases proportionately with the internal energy of the ion before reaching a maximum plateau beyond which no rate enhancement is observed. The shape of the curve from a plot of k versus energy E is predominately determined by the geometry of the transition state and the value of E_0. Use of the QET allows the rate of fragmentation of an ion with a given internal energy to be predicted. The maximum rate constant for a simple bond cleavage process is of the order of 10^{14} s^{-1}.

A simplified version of the QET enables the abundances of molecular and fragment ions in mass spectra to be described in a semi-quantitative manner without the aid of a computer. In this version, the rate of fragmentation of an ion (k) is given by equation 2.7 where v is a frequency factor influenced by the entropy of the process and N the number of oscillations (rotations or vibrations) possible.

$$k(E) = v[(E - E_0)/E]^{N-1} \tag{2.7}$$

As explained above, the appearance of fragment ions in a mass spectrum is ultimately influenced by the time allowed for such fragmentation processes. Ions spend approximately 10^{-6} s (or one microsecond) in an electron ionisation source so that only relatively fast fragmentation processes occur in this region of the mass spectrometer. On some mass spectrometers, an entire mass spectrum is recorded in just a few hundred microseconds providing energetically-excited ions with only this amount of total time to fragment.

2.2.5 Metastable Ions

Metastable ions, denoted m^*, are those formed by unimolecular dissociation of molecular ions in the field-free regions of the mass spectrometer anywhere between the ion source and detector. They can be useful in establishing the fragmentation pathways of molecular ions by unequivocally linking a fragment ion with a specific precursor.

Consider a molecular ion of mass m_p that dissociates to a fragment ion plus a radical or neutral molecule (equations 2.3 and 2.4). If a metastable fragment ion is produced in flight from the ion source to the detector it will have less translational energy than a comparable fragment ion generated within the ion source (of mass m_f). This results in the ions appearing in the mass spectrum at an apparent mass (m^*) that is less than that of fragments generated in the source (m_f). On a magnetic sector mass spectrometer, the mass of a metastable ion, m^*, is expressed according to equation 2.8.

$$m^* = m_f^2/m_p \tag{2.8}$$

The widths of the ion signals of these metastable ions of mass m^* are considerably greater than those of other ion fragments. This is because the ions are not energy focused as they leave the ion source and the kinetic energy of the precursor ions is released *isotropically* as they are formed. Thus metastable ions have a broader range of energies over other fragment ions.

As an example, the loss of a methyl radical from ionised propane (of a molecular weight of 44 Da) gives rise to a fragment ion at m/z 29 and a metastable ion at m/z 19.1 ($29^2/44 = 19.1$) Because the metastable ions appear as broad peaks in the mass spectrum, their mass-to-charge ratios are usually quoted to only a few decimal points at most.

2.3 RELATIVE ION ABUNDANCE

The appearance of an ion in a mass spectrum is the result of an electrical current that is generated and amplified when the ion strikes the detector. A measure of the ion current across all ions in a mass spectrum is referred to as a *total ion current* (TIC). The charge on a singly charged ion is 1.6×10^{-19} coulombs (C). Since one ampere represents one coulomb of charge per second, when one million singly-charged ions strike the detector a current of 1.6×10^{-13} amperes (A) will be produced.

The vertical or y-axis of a mass spectrum is usually plotted to display the ion current, the number of ions, or more commonly *relative ion abundances*. Here the ion signal with the highest current is normalized to 100% on the relative abundance scale and all other ions have abundances measured relative to this peak, the *base peak*. The relative abundance of an ion is dependent on a number of factors including the stability of the ion, the stability of the neutral product (in the case of a fragment ion from a unimolecular decomposition), suppression effects, mass resolution, and detector efficiency.

Ion detectors do not detect ions across the m/z scale with equal efficiency, and it is common for ions at high m/z to be detected less readily than those at low m/z. Mass resolution also impacts ion abundance measurements. An ion signal associated with two unresolved ions would have a higher intensity reflecting the contribution of both ion currents to the signal than would be the case if each were mass resolved. Suppression effects have also been observed widely in mass spectrometry. In some cases, a neighbouring ion can completely suppress the signal of another ion such that it is not detected. When ionised separately, both ions are easily detected.

Entropy effects are an important consideration in terms of the intensity or abundance of fragment ions formed by unimolecular dissociation. Entropy considerations favour the production of fragment ions by simple cleavage reactions over rearrangements involving a change in the molecular structure.

2.4 MASS RESOLUTION

The ability to separate two ion signals from one another in a mass spectrum is defined by *mass resolution* (R). Traditionally, this measure has been made based on the resolution of two ion signals above 10% of their height (the so-called 10% valley definition) (Figure 2.3). Mass resolution is defined by equation 2.9 where M_1 represents the *m/z* ratio for the first ion and M_2 represents the *m/z* ratio of the second, and $M_1 > M_2$.

$$R = M / \Delta M = M_1/(M_1 - M_2) \tag{2.9}$$

Therefore in order to resolve ions at *m/z* 1000 and 1001 with a 10% valley, a mass resolution of 1000 is required. The same is true for ions at *m/z* 500 and 500.5. In some cases, a mass resolution is referred to as a ratio, namely 1:1000.

Mass resolutions vary due to a number of factors including the *initial kinetic energy* of the ions as they exit the ion source. Most ions of the same *m/z* are formed with a range of initial energies (often represented by a bell-shape distribution), a result of their proximity to or remoteness from the acceleration lenses to which high potentials are applied. This may be corrected for by the mass analyser by what is known as *energy focusing*. Here only ions of the same *m/z* and energy are allowed to pass to the detector but this is achieved at the expense of ion detection. In cases where this energy spread is not corrected for, the width of the ion signal recorded at the detector will be larger.

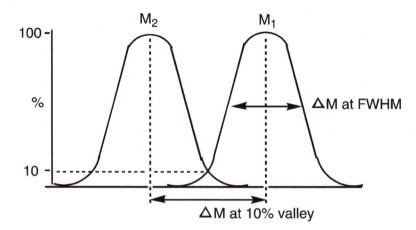

Figure 2.3 *Definition of mass resolution at a 10% valley and full-width at half maximum (FWHM)*

The most important factor, however, that impacts mass resolution is the type of mass analyser used. Magnet-based instruments as a rule achieve superior mass resolution to quadrupole and time-of-flight mass analysers. However, a better understanding of ion motion and optics inside a mass spectrometer has led to improvements in instrument design that have considerably increased the mass resolutions that can be achieved on most instruments. Mass resolutions of the order of 100–500 are considered low, those of the order of 1000–5000 are considered moderate, while those above 10,000 are considered to be high (see Appendix 4). Fourier transform ion cyclotron resonance (FT-ICR) mass spectrometers, described in the next chapter, offer the highest mass resolutions attainable to date, routinely up to 100,000. The ion signals associated with isotopes, even for large macromolecules, can thus be resolved to baseline (or the x-axis) on these instruments.

Where isotopes or components in mixtures are not mass resolved, some measurement of mass resolution is often desired. In this case, it is common to measure mass resolution based upon the width of the ion signal at half its height. This measure is referred as *full width at half maximum* (FWHM) (Figure 2.2). For an ion at *m/z* 1000, with a peak width of 2u at half its height, the mass resolution is said to be 500 (or 1000/2).

2.5 MASS MEASUREMENT AND ACCURACY

The capacity of mass spectrometers to resolve ions associated with heavy isotopes (*e.g.* ^{13}C, ^{2}H) from those that contain only the lightest isotopes, enables a mass measurement to be made based on any one or several of these ion signals. For small molecules, the ion signals associated with heavy isotopes are typically of a low relative abundance since such isotopes are rare in nature. As a result, most mass measurements are based on the *peak of lightest isotopes*, and the resulting molecular weight of the compound is referred to as its *monoisotopic mass*. For carbon-based compounds, this is also referred to as the *^{12}C-only mass*, since ion signals at the lowest *m/z* ratios in the isotopic distribution contain no ^{13}C or ^{14}C. Carbon is considered exclusively here since the natural occurrence of heavy isotopes for other common elements in organic compounds are much lower compared with carbon.

Where an ion's isotopes are unresolved, a molecular weight measurement is based upon the *m/z* value at the peak top or a centroid value (where the centre of the peak top is determined from the area above a particular ion intensity level). Such a measurement yields an *average mass*, which contains a contribution to the molecular weight of the compound from all elemental isotopes. By definition the average molecular weight of a compound is always greater than its monoisotopic value. This difference between the monoisotopic and average value can be quite significant for large molecules. As an example, the monoisotopic molecular weight of the protein ubiquitin, with an elemental formula of $C_{378}H_{629}N_{105}O_{118}S_1$ has a calculated value of 8,559.62 and average value of 8,564.86 Da. Special care must be taken where mass measurements are made based upon the calibration of the mass-to-charge scale using some ion signals with isotopes resolved and others where they are not.

As the molecular weight of a compound becomes large, the probability that any of its ions contain no heavy isotopes becomes exceedingly small. As a result, the peak of lightest isotopes for any of its ions becomes small relative to those ions that contain heavy isotopes. It is necessary in these circumstances to base the molecular weight on the *m/z* value of ions containing some level of heavy isotope. In the case of the electrospray mass spectrum of bovine ubiquitin, a mass measurement based on the ion signal for ions containing five ^{13}C atoms (or some other combination of heavy isotopes of equal mass, *e.g.* $^{13}C_4 + {}^{2}H$) at *m/z* 779.613, is 8,564.66 (11 × 779.613 – 11 × 1.0078 to adjust for the 11 protons attached to the protein that gives the ions their charge). This experimental value is within 0.03 u of the theoretical value (8,564.63) at this level of isotope enrichment. Hence a mass accuracy of $0.04/8,564.63 = 3.5 \times 10^{-6}$ or 3.5 parts-per-million (ppm) has been achieved.

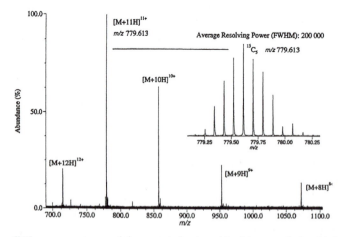

Figure 2.4   *ESI mass spectrum of the protein bovine ubiquitin recorded at high mass resolution (~200,000 at FWHM). Insert shows resolved isotope peaks for the [M + 11H]$^{11+}$ ions.*

The difference in mass-to-charge ratio between the isotope peaks for any ion (shown enlarged for the $[M + 11H]^{11+}$ ions to be 0.09 or 1/11 in Figure 2.4) is $1/z$. Thus the charge on any ion can be determined directly from the mass spectrum where its isotopes are mass resolved.

The level of heavy isotope content must be known or predicted in order for the molecular weight of the compound not to be in error. Alternatively, the compound can be constructed in an environment depleted of heavy isotopes such that ions associated with only the lightest isotopes for all elements are detected.

FURTHER READING

F.W. McLafferty and F. Turecek, *Interpretation of Mass Spectra*, 4th edition, University Science Books, 1993.

E. DeHoffmann and J. Charette, *Mass Spectrometry – Principles and Applications*, 2nd edition, John Wiley and Sons, New York, 2001.

H.M. Rosenstock, M.B. Wallenstein, A.L. Wahrhaftig and H. Eyring, Absolute Rate Theory for Isolated Systems and the Mass Spectra of Polyatomic Systems, *Proc. Natl. Acad. Sci. U.S.A.* 1952, **38**, 667–678.

R.G. Cooks, J.H. Beynon, R.M. Caprioli and G.R. Lester, *Metastable Ions*, Elsevier, Amsterdam, 1973.

CHAPTER 3

The Mass Spectrometer

3.1 BASIC COMPONENTS

All mass spectrometers consist of three basic components, the:

(i)   *ion source,*
(ii) *mass analyser,*
(iii) *ion detector.*

The role of the *ion source* is to introduce molecules into the mass spectrometer and convert them to a charged or ionised form. The ion source like the rest of the mass spectrometer is usually, though not always, held at a low pressure. Mass spectrometers are operated under vacuum to prevent the collision of ions with residual gas molecules during their flight from the ion source to the detector. This is because the ions are formed with excess energy and this, together with their charged character, can result in their reaction with other gaseous material present. To avoid this, the levels of contaminants and atmospheric gases such as oxygen within the ion source should be minimised. The ideal operating pressure is that in which the average distance an ion travels before colliding with a gas molecule (its *mean free path*) is longer than the distance from the source to the detector.

After ions are formed in the source, they are accelerated into the *mass analyser* where they are separated *in vacuo* according to their mass and charge through the use of electric and/or magnetic fields. Finally, the ions are passed onto an *ion detector* producing an electrical current that is amplified and detected.

In most mass spectrometers, these three basic components are physically discrete entities. Thus each of them will be considered separately in order to understand how ions are formed, separated and detected in mass spectrometry experiments. It is not possible to review all the ionisation techniques that have been used in mass spectrometry experiments, some of which have been replaced by other more efficient methods.

Instead, the following section discusses those most widely used approaches either alone, or in conjunction with, chromatographic and electrophoretic separations.

3.2 IONISATION TECHNIQUES AND INTERFACES

3.2.1 Electron Ionisation

As described in Section 2.1, the traditional method of ion production in mass spectrometry is *electron impact* or *electron ionisation* (EI) in which gaseous sample molecules are bombarded with a stream of electrons (equation 2.1). Other processes by which ions can be formed during electron ionisation include *dissociative ionisation* (equation 3.1), *ion pair formation* (equation 3.2) and *electron capture* in which negatively charged ions are produced (equation 2.2). The electron ionisation technique is widely used for the study of relatively volatile organic molecules by mass spectrometry.

$$AB + e^- \rightarrow A^{+\cdot} + B + 2e^- \tag{3.1}$$

$$AB + e^- \rightarrow A^+ + B^- + e^- \tag{3.2}$$

The electrons are produced by heating and passing a current through a thin ribbon of metal (such as ruthenium) known as the filament or cathode. The electrons are projected across the ion source by their attraction to an anode on the opposite side of the chamber (Figure 3.1). The energy of the electrons depends on the difference in the potentials applied to the cathode and anode. If a voltage difference of 70 V is maintained, the electrons have energies of 70 electron volts (eV), or the equivalent of 6.8×10^3 kJ mol^{-1}. A small magnetic field is usually applied across the ion source to cause the electrons to follow a helical path in order to increase the probability that they interact with a gaseous sample molecule. A pressure of typically 10^{-6} Torr or 1.3×10^{-4} Pa is maintained in the ion source.

The collision of electrons with the sample molecules, M, often leads to their ionisation by electron loss with the formation of the molecular ion, $M^{+\cdot}$ (equation 2.1). The energy required to ionise a molecule depends on the molecule itself but, as an example, an energy of 10.5 eV is needed to ionise a molecule of ethanol. Most organic molecules ionise in the range of 8–15 eV. Yet since not all the electron energy is necessarily transferred to the sample molecules during collision, higher electron energies of ~70 eV are commonly used in EI mass spectrometry.

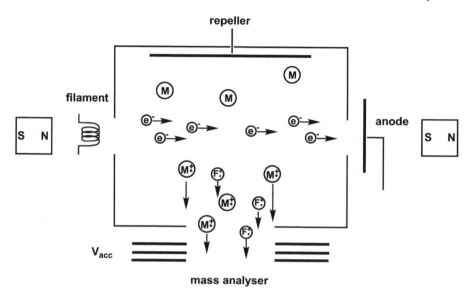

Figure 3.1 *Schematic representation of an electron ionisation source*

Regardless of the energy of the electrons, the sample molecules will possess a range of energies up to and including the energy of the electrons. Therefore some molecules will have sufficient energy to ionise (equations 2.1 and 2.2), others will not, and still others will contain enough energy to dissociate (equations 2.3, 2.4, 3.1, 3.2). Several factors contribute to the formation of fragment ions including the strength of the bonds to be broken, the stability of the products of fragmentation (both the ions and neutrals or radicals), and the internal energy of the fragment ions themselves. Where the dissociation of the molecular ions is problematic (and as shown in Chapter 5 it is often useful to determine the structure of a molecule), a lower potential difference (10–15 V) can be applied between the filament and the anode.

The voltage difference at which molecular ions are first observed in a mass spectrometer is known as the *ionisation potential*. A few volts above this, molecular ions are mostly formed. As the potential increases further, more fragment ions are produced (Figure 3.2).

In principle, mass spectrometers can be used to measure ionisation potentials. Because these measurements can be unreliable, a better measure is the efficiency of producing fragments. The *appearance potential* is related to the overall energy for the processes shown in equations 2.1, 2.3 and 2.4. This information is useful for measuring the thermodynamic parameters of a molecule such as its heat of formation and bond dissociation energies (see Chapter 6).

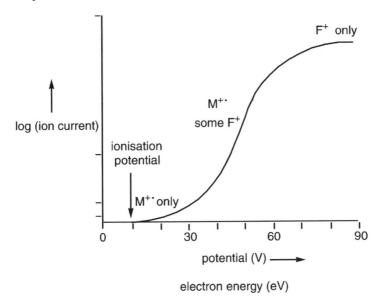

Figure 3.2 *Plot of ion current versus the potential difference between the filament (cathode) and anode. The ionisation potential represents the voltage at which molecular ions are first detected*

Since electron ionisation can lead to the production of fragments (F^+) (equation 2.3) as well as intact molecular ions, it is referred to as a *hard ionisation* method. The remaining ionisation methods discussed in this chapter result predominantly in the formation of ions without fragmentation, and hence are known as *soft ionisation* techniques.

3.2.2 Chemical Ionisation

Chemical ionisation (CI) is related to the electron impact method except that ionisation of a reagent gas, rather than the sample molecule itself, occurs first. This is followed by the transfer of charge to the sample molecule by a chemical process. One of the most common reagent gases is methane. When subjected to electron impact, a molecule of methane can ionise to form $CH_4^{+\bullet}$ by electron loss. This ion can react with a second molecule of methane, to produce CH_5^+ (equation 3.3).

$$CH_4^{+\bullet} + CH_4 \rightarrow CH_5^+ + CH_3^{\bullet} \qquad (3.3)$$

The ion CH_5^+ is an efficient proton donor, so that a sample molecule M also present in the ionisation chamber can be ionised according to equation 3.4.

$$M + CH_5^+ \rightarrow [M + H]^+ + CH_4 \qquad (3.4)$$

To prevent the direct ionisation of molecules M, methane is present in the ion source at a much higher concentration than the sample. Because of this, a chemical ionisation source operates at a much higher pressure (10^{-3} to 10^{-4} Torr, or 0.1 to 1 Pa) than an EI source.

There are many chemical processes other than proton transfer that can be achieved inside an ion source to effect chemical ionisation. These include *charge transfer* (or *charge exchange*) (equation 3.5), ion-molecule addition (equation 3.6), and even nucleophilic displacement reactions in the case of negatively-charged ions (equation 3.7).

$$M^{+\bullet} + N \rightarrow M + N^{+\bullet} \tag{3.5}$$

$$M^{+} + N \rightarrow [M + N]^{+} \tag{3.6}$$

$$Nu^{-} + AB \rightarrow Nu\text{-}A + B^{-} \tag{3.7}$$

3.2.3 Coupling Gas Chromatography to Mass Spectrometry (GC-MS)

The ability to ionise volatile molecules within a mass spectrometer led to mass spectrometers being coupled to gas chromatographs (GC). In these experiments, the volatile components of relatively complex sample mixtures can be ionised (by either electron or chemical ionisation) and detected in a stepwise manner as they are released from the GC-column. Modern GC-MS mass spectrometers use capillary columns (with 100–500 µm internal diameters) that provide for the separation of low levels of analytes and also minimise the amounts of GC carrier gas that enter the ion source. This helps to maintain a low operating pressure in the source. Modern GC-MS interfaces are designed to further minimise the levels of carrier gas that enter the ion source by using high capacity pumping systems. However, it is a requirement that some compromise be made between optimal GC conditions and those required for MS operation. A so-called *open-split interface* in which a proportion of the eluent from the column is pumped away before it enters the ion source is one way to achieve the coupling of a GC and mass spectrometer. This interface also enables the GC column to be interchanged without breaking vacuum ("venting") to the mass spectrometer.

3.2.4 Field and Plasma Desorption Ionisation

These two ionisation techniques have largely been superseded by other methods, but they are mentioned briefly here since they represent the first methods available for ionising non-volatile molecules. Field desorption (FD) ionisation is achieved by depositing the sample onto a metal filament coated with carbon. A potential difference is applied between the filament and a nearby electrode such that ions are desorbed from the surface. The method is particularly useful for large non-polar compounds such as hydrocarbons but requires some skill to correctly prepare the coated filament.

In plasma desorption (PD) the sample is deposited onto a foil constructed of nickel or aluminium-coated nylon. The fission fragments from the radioactive decay of californium-252 then pass through the foil. Californium-252, an isotope with a half-life of about 2.6 years, is a very efficient neutron source with one microgram producing about 170 million neutrons every minute. The fission particles deposit considerable energy into the sample and lead to the direct release of ionised forms of the sample molecules, usually $[M + H]^+$ and $[M + Na]^+$ ions.

Plasma desorption (PD) ionisation was the first ionisation technique capable of ionising non-volatile sample molecules with molecular weights of the order of 10,000 Da including polar molecules such as proteins. However, the use of a radioactive source and difficulties with preparing the sample for ionisation led to it being largely replaced by fast atom bombardment (FAB) ionisation shortly after its discovery.

3.2.5 Fast Atom or Ion Bombardment

Fast atom bombardment (FAB) is related to an approach widely used to study the chemical nature of materials and surfaces called *secondary ion mass spectrometry* (SIMS) (see Chapter 9). In SIMS, a primary beam of high-energy (typically 10–30 keV) ions such as Xe^+ or Cs^+ bombard a solid surface, releasing secondary sample ions for analysis. One problem with the approach is that once the sample is ablated from a position on the surface, the yield of secondary ions decreases unless the primary ions are made to strike a different position.

To overcome this, Michael Barber and colleagues invented the FAB approach in which the sample compound was suspended or dissolved in a non-volatile viscous liquid. Once a portion of the sample is ablated or sputtered from the surface by a primary beam of atoms or ions, the liquid "matrix" flows back across this region restoring a proportion of sample (Figure 3.3). The matrix also serves an additional role by dissipating the energy from the primary beam to minimise molecular damage to the sample.

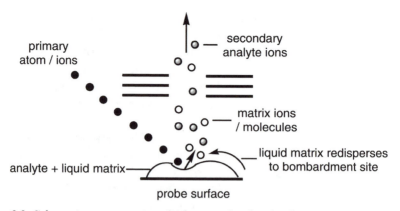

Figure 3.3 *Schematic representation of a fast atom/ion bombardment (FAB) source*

The original application of the FAB technique made use of high-energy atoms (rather than ions) of argon to bombard the sample dissolved in glycerol. These atoms are formed from a gas held at a relatively high pressure (10^{-3} to 10^{-4} Torr, or 0.1 to 1 Pa) in an *atom gun* by a charge exchange process with ionised gas (equations 3.8 and 3.9).

$$Ar + e^- \rightarrow Ar^{+\bullet} + 2e^- \tag{3.8}$$

$$Ar^{+\bullet} \text{ (fast)} + Ar \text{ (slow)} \rightarrow Ar \text{ (fast)} + Ar^{+\bullet} \text{ (slow)} \tag{3.9}$$

When the primary atoms or ions collide with the liquid surface, a charge is transferred to the involatile liquid matrix (M) molecules and subsequently to the analyte (A) sample molecules (equations 3.10–3.12).

$$Ar^{+\bullet} + M \rightarrow M^{+\bullet} + Ar \tag{3.10}$$

$$M^{+\bullet} + MH \rightarrow MH^+ + M^\bullet \tag{3.11}$$

$$MH^+ + A \rightarrow AH^+ + M \tag{3.12}$$

Thus secondary ions are produced from the liquid matrix ($M^{+\bullet}$ and MH^+) as well as the sample analyte (usually AH^+). Because the concentration of the former is much higher on the probe surface, FAB spectra have a characteristic high "background" at low to modest m/z ratios (up to 500 u) associated with ions of individual matrix molecules and their molecular clusters. This matrix background can obscure the detection of analyte ions below m/z 500 particularly where the levels of sample present are low. A representative early FAB mass spectrum for the peptide met-lys-bradykinin (MW 1318 Da) is shown in Figure 3.4.

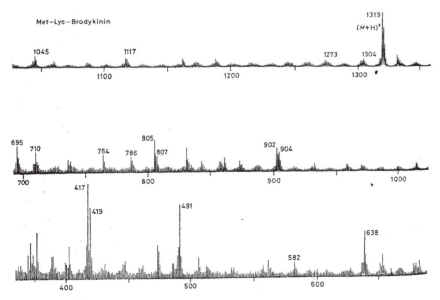

Figure 3.4 *FAB mass spectrum of the peptide met-lys-bradykinin*
(Source: M. Barber, R.S. Bordoli, G.J. Elliott, A.N. Tyler, J.C. Bill and
B.N. Green. Fast atom bombardment (FAB) mass spectrometry: A new ion
source for mass spectrometry, *J. Chem. Soc. Chem. Commun.*, 1981, 325–327,
Figure 1)

Several properties of the liquid matrix are important to the success
of the FAB method. The matrix should be an involatile liquid in vacuum, it should be chemically inert, and preferably dissolve most analytes. FAB matrices often possess a low *pKa* to assist in the generation
of positively-charged AH^+ ions, or a high *pKa* where negatively-
charged ions are to be produced. Glycerol, nitrobenzyl alcohol,
thioglycerol and dimethylsulphoxide are common FAB matrices. Dithiothreitol and dithioerythritol can be added to the matrix thioglycerol
(*magic bullet*) to reduce disulphide bonds in proteins suspended or
dissolved in the matrix. Ions are typically produced by FAB ionisation
over a period of several minutes before it is necessary to replenish the
liquid matrix and/or the sample. Primary ions such as Cs^+ have largely
replaced high-energy argon atoms in most FAB experiments, and thus
the experiments have been referred to as liquid SIMS (LSIMS) in some
accounts.

Samples can also be introduced directly from a high-pressure liquid
chromatograph (HPLC) by means of continuous flow FAB. This technique involves adding the FAB matrix to the mobile liquid phase used in
the HPLC experiment at approximately 5% by volume. The solution is
pumped down the length of the FAB inlet probe onto a target (Figure

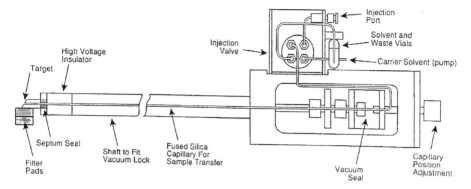

Figure 3.5 *Design of a continuous flow CF-FAB sample probe*
(Source: R.M. Caprioli and W.T. Moore, in *Methods in Enzymology*,
McCloskey (ed), Academic Press, New York, 1990, Vol. 193, Ch. 9, p. 216,
Figure 1)

3.5) or mesh frit. A filter below the probe tip is used to absorb excess
matrix. Stable ion signals are produced when the amount of liquid
delivered to the tip or frit is equal to the rate of evaporation. The reduced
concentration of matrix on the tip over standard or static FAB leads to a
lower background of matrix ions that can aid the detection of analyte ions.

Although FAB ionisation is less practiced today due to the development of laser based and spray ionisation methods, it remains an important technique for many applications. It is still in use for the study of
organic and smaller biological compounds in many laboratories.

3.2.6 Laser Desorption and MALDI

Lasers operating in both the ultraviolet (UV) and infrared (IR) have been
used to desorb and ionise samples from solid surfaces for some time. The
transfer of energy from the laser pulse leads to electronic excitation of
the sample. Laser powers vary from approximately 10^6 to 10^{10} J sec^{-1} cm^{-2}
where the total energy per pulse is of the order of a few millijoules to a
joule. Apart from the region of the sample upon which the laser is
focused, there is usually little excess energy dissipated through the
analyte. That said, considerable decomposition of some analytes can
occur following the laser pulse. Despite this, a number of moderately-
sized (~1000 Da) sample molecules such as oligosaccharides, peptides
and polymers have been successfully ionised by laser desorption ionisation (LDI, or just LD). This "neat" desorption strategy was effectively
replaced with the development of *matrix-assisted laser desorption
ionisation* (MALDI).

In MALDI, the analyte of interest is mixed with a large mole excess of (*ca.* 1,000-fold) a matrix compound that absorbs efficiently at the laser wavelength. The matrix allows the energy from the laser to be dissipated and also assists with the ionisation sample molecules through electron transfer and chemical processes. Both solid and liquid matrices have been used, though the former is by far the more successful in most applications.

Where a UV laser is used, common MALDI matrices include nicotinic acid, 2,5-dihydroxybenzoic acid, sinapinic acid and α-cyano-4-hydroxy-cinnamic acid (Figure 3.6). It is important to note that all of the MALDI matrices in Figure 3.6 contain phenolic and/or carboxylic acid groups. It has been found that proton transfer from matrix to sample molecules is important in order to achieve efficient ionisation of many compounds and that this transfer occurs, at least in part, in the vapour phase above the sample plate.

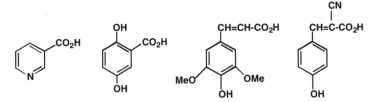

Figure 3.6 *MALDI matrices nicotinic acid, 2,5-dihydroxybenzoic acid, sinapinic acid and α-cyano-4-hydroxycinnamic acid (in order from left to right)*

The correct preparation and deposition of MALDI samples onto the sample plate is critical to the success of the method. Most sample solutions are diluted in a solution of matrix, and a small volume of the combined solution (1 μl or less) is deposited onto the sample plate. The sample droplet is usually allowed to dry in ambient air (by so-called *dried droplet* evaporation). However, since the morphology of the crystallised sample surface affects the success of MALDI mass spectrometry, other methods to deposit samples have been developed. These include the addition of organic solvents to the solution, the use of heat to assist the drying process, and the electrostatic-spraying of solutions of analyte and matrix (either separately or in a combined form). The latter technique gives rise to extremely thin and uniform surface of both the analyte and matrix resulting in more reproducible mass spectra being obtained regardless of the position from which the sample is ablated by the laser.

A further useful strategy that has been adopted where the concentrations of analyte are low is the deposition of droplets onto pre-coated sample surfaces to either localise (for example, through the use of

Teflon-based surfaces) or immobilise the analyte molecules onto a very small sample area (<1 mm²). Subsequent chemistries can also be performed on the molecules on these surfaces prior to their analysis by mass spectrometry. One such MALDI-based approach has been dubbed SELDI for *surface-enhanced laser desorption ionisation.*

Since the laser or sample surface can be easily repositioned during analysis, most MALDI-based mass spectrometers today make use of a sample stage onto which 100 to several hundred samples are loaded. The plates also resemble the size of a microgel or blot (some 10 cm²) to facilitate the direct transfer of samples (particularly proteins) after their separation by two-dimensional polyacrylamide gel electrophoresis (2D-PAGE). Greater success has been achieved in experiments where the proteins are blotted onto a membrane to which a solution of matrix is applied. Unfortunately, the direct ionisation of proteins from gels has proved more difficult due to the presence of detergents and other contaminants that impede the ionisation process.

A representative MALDI mass spectrum for a protein mixture is shown in Figure 3.7. Consistent with most protein analyses, the dominate ions in the spectrum have the form [M + H]⁺ from which a molecular weight can easily be derived. Note in this case that insufficient mass resolution has been obtained in this case to resolve the isotopes for the [M + H]⁺ ions.

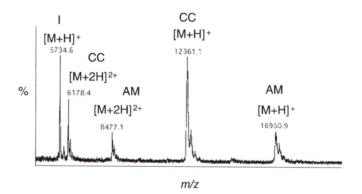

Figure 3.7 *Linear MALDI-TOF mass spectrum of the protein mixture, insulin (I), cytochrome c (CC) and apomyoglobin (AM)*

MALDI is now a firmly established technique, particularly for the study of polar, high molecular weight compounds such as proteins, glycoconjugates and nucleic acids. In most mass spectrometry laboratories today, its use is complemented by the application of *electrospray ionisation* (ESI), a completely unrelated ionisation method described in the next section.

3.2.7 Spray Ionisation Methods; Thermospray

One disadvantage of MALDI, and other ionisation methods, in which the sample is deposited in its solid state is the difficulty in performing high throughput separations in conjunction with mass spectrometric analysis. Although mass spectrometry can be used to analyse sample mixtures directly, some components may not be mass resolved and thus not be detected as the complexity of these mixtures increases.

Various chromatographic and electrophoretic approaches are used widely for the separation of components within complex chemical and biological extracts. The development of solution-based spray ionisation approaches have enabled these technologies to be coupled directly to a mass spectrometer providing a further separation dimension prior to MS analysis.

In 1983, Blakley and Vestal reported the development of the *thermospray ionisation* method for this purpose. Briefly, a dilute solution of an analyte is pumped through a stainless steel tube and heated to approximately 100 °C subject to the flow rate and nature of the solution. A jet of vapour containing a mist of solution droplets is projected into the ion source by a free jet expansion and preformed ions in solution are evaporated and detected. Thermospray ionisation has been applied to the study of small organic molecules and moderately-sized biological molecules such as peptides but has been largely superseded by electrospray ionisation.

3.2.8 Electrospray Ionisation

Electrospray ionisation (ESI) was first conceived in the late 1960s by Malcolm Dole and has developed from experiments performed in the late 1980s by John Fenn and colleagues. The electrostatic spraying of liquids is used in many industrial applications and involves passing a solution through a needle held at high voltage (typically 4–5 kV) relative to some counter electrode. When the solution is an electrolyte and the needle forms part of an ion source in a mass spectrometer, the fine mist of droplets that emerge from the needle tip possess a net positive or negative charge determined by the polarity of the needle and are attracted to the entrance of a mass analyser (Figure 3.8). The droplets emerge from what is known as a *Taylor cone* formed by the elongation of the electrolyte solution at the needle tip as like-charged ions are repelled from the needle. The application of a "counter-current" dry gas (that passes in the opposite direction to the passage of droplets) considerably aids droplet evaporation and it was this feature that led to the successful ionisation of large molecules.

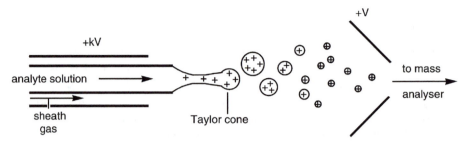

Figure 3.8 *Magnified representation of the cross-section of an electrospray ion source.*

As the droplets evaporate, the ions within them move closer together. At some point there are sufficient Coulombic repulsive forces between the ions to overcome the liquid surface tensions, resulting in the production of smaller droplets that continue to undergo the process. Eventually solvent-free ions are produced that are passed through the mass analyser and detected.

A curious feature of the electrospray ionisation mass spectra of large biopolymers is that the ions produced are usually multiply-charged, with a continuous series of such ions being detected (Figure 3.9). The reasons for this are not entirely clear though the phenomenon has been linked to the time it takes for ions to emerge from the solution droplets during the evaporation process.

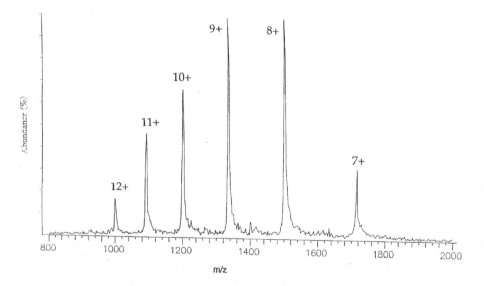

Figure 3.9 *Electrospray ionization (ESI) mass spectrum of protein chloroflexus thioredoxin exhibiting ions of the form $[M+nH]^{n+}$*
(Source: K.M. Downard, Advances in protein analysis and sequencing by mass spectrometry, *New Adv. Anal. Chem.*, 2002, **P2**,1–30, Figure 1 (adapted))

An advantage of this feature is that a molecular weight measurement can be made based on each multiply-charged ion and the values averaged across all charge states. This leads to routine mass accuracies of $\pm$ 0.01%, or a 1 Da error at a molecular weight of 10,000 Da even on mass spectrometers with modest mass resolving capabilities. To achieve this, the charge state for a particular ion must be determined. If it is assumed that two neighbouring ions (with mass-to-charge ratios of m_i/z_i and m_j/z_j where $m_i/z_i < m_j/z_j$) differ by one charge unit and both support the same charge-bearing species, then the charge on ion i (z_i) is given by equation 3.13 where m_p is the mass of the charge-bearing ion, typically a proton.

$$z_i = (m_j/z_j - m_p)/(m_j/z_j - m_i/z_i) \qquad (3.13)$$

Once the charge of any ion is derived, a molecular weight measurement based on any ion signal can be determined from equation 3.14.

$$MW = z_i\,(m_i/z_i) - z_i\,m_p \qquad (3.14)$$

Applying this equation in the case of the data presented in Figure 3.9, the charge of the protonated ions at m/z 1,335.4 is $z = (1,502.2-1)/(1,502.2-1,335.4) = 9.0$. The molecular weight of the protein based on this protonated ion signal is thus $9(1,335.4) - 9(1.0) = 12,009.6$. The same molecular weight value is obtained based on the m/z for the $[M + 8H]^{8+}$ ion; that is $8(1,502.2) - 8(1.0) = 12,009.6$.

Several computer algorithms have been developed to perform these calculations automatically and to display the output on a molecular weight scale in what has become known as a *deconvoluted mass spectrum*. One such algorithm considers the mass of the charge-bearing ions as a variable and constructs a three-dimensional deconvoluted mass spectrum from which the identity of the charge-bearing species can be determined and not assumed.

3.2.9 Atmospheric Pressure Chemical Ionisation

A related approach that is also capable of ionising polar molecules directly from solution is known as *atmospheric pressure chemical ionisation* (APCI). In this method, ionisation is achieved by an electrical discharge in the vicinity of gaseous sample molecules produced by vapourising the solution stream. Ionisation takes place through chemical processes such as those described in Section 3.2.2. The process is less efficient at ionising large molecular weight compounds, but does have utility for many modestly-sized polar biomolecules (to ~ 1000 Da). For this reason, the approach is employed widely in drug discovery investigations including the study of metabolites (see Chapter 8). It works well even for relatively high flow rates (several ml min^{-1}) that are common in analytical HPLC applications. Since the ESI and APCI ionisation approaches are related and have complementary applications, many instruments feature interchangeable ESI and APCI ion sources.

A disadvantage of ESI mass spectra of large compounds is that many ions are associated with each component present in the sample solution. In the case of a complex sample mixture, these ion distributions could be incorrectly associated with one another so that molecular weight errors arise. However, to overcome this the ionisation method can be coupled to a liquid chromatograph or capillary electrophoresis system where some initial separation of components is effected prior to MS analysis.

3.2.10 Coupling Liquid Chromatography and Capillary Electrophoresis with Mass Spectrometry

There are two major considerations in coupling liquid chromatography and capillary electrophoresis separation systems to a mass spectrometer. First, the solvent must be efficiently evaporated before ions leave the source to enable the pressure within the mass analyser to be maintained. This also aids the efficient sampling of ions. Second, the nature of the solvent and other dissolved components (buffers, salts, denaturants *etc.*) should not impede the ionisation process. There is clearly a compromise reached in order to optimise the separation of chemical and biological mixtures with their direct on-line detection by mass spectrometry. Nonetheless, reproducible data can be acquired when appropriate operating conditions are maintained.

A typical liquid chromatographic ESI mass spectrometry (LC-ESI-MS) apparatus is illustrated in Figure 3.10. In brief, solvent pumped from the reservoirs enters the injector and passes through the chromatography column to the mass spectrometer. A UV detector can also be incorporated such that components are detected by absorption spectroscopy and mass spectrometry either simultaneously or in tandem. Once the sample is injected into the loop, the solvent delivers it to the column where chromatographic separation of the components occurs. The components then pass one by one, or as simpler mixtures, into the ion source where they are ionised and ultimately detected. Note that since the mass spectrometer separates the ionic forms by mass analysis, it is not

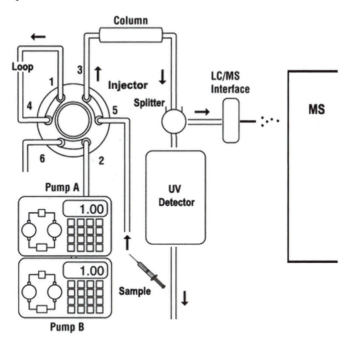

Figure 3.10 *Representation of a liquid chromatographic ESI mass spectrometry (LC-ESI-MS) experiment*

necessary that complete chromatographic separation of each component of a mixture be achieved. Thus the chromatographic separation step is less rigorous than that which would be required where no mass spectrometer was employed.

In the case of liquid chromatography, solvent flow rates of the order of several nanolitres to millilitres per minute have been coupled to ESI and APCI sources. An approximately linear correlation between the yield of ions detected and the concentration of the analyte in solution is observed over several orders of magnitude. Thus a quantitative measure of the ion current for each component enables the relative concentration of the analytes in solution to be determined. The addition of standards of known concentration into the sample, or their analysis in a separate run, allows such LC–MS approaches to be used for quantitative applications.

The preferred mobile phases are those that contain significant levels of a volatile organic solvent such as acetonitrile or methanol, both of which are widely used for chromatographic separations. Pure aqueous solvent systems, however, can also be managed albeit with slightly reduced performance. The presence of low levels of ion pairing agents such as trifluoroacetic acid can assist with generating preformed ions of the sample components in solution. High levels of salt and other buffers and denaturants, however, should be diverted away from the ion source, usually during the early stages of the run, in order to prevent their build up in the transfer lines and on the spray needle.

One further issue in performing such experiments on most mass spectrometers is whether or not all ions across the *m/z* range of the mass analyser are detected simultaneously. Where mass analysers are used in which the electric and/or magnetic fields are scanned, ions produced from an analyte eluting over a relatively small time period may pass into the mass analyser but not be transmitted to the detector. To compensate (if not correct) for this, either fast scanning of the mass analyser is employed or the flow rate of the mobile phase is reduced.

An alternate method to detect low levels of compounds in these experiments is to "park" the electric and/or magnetic fields of the mass analyser to transmit only ions of a particular *m/z* ratio onto the detector. This *selected ion monitoring* (SIM) mode is useful for quantitative analysis of particular components in chemical and biological samples.

Capillary electrophoresis mass spectrometry (CE-MS) is achieved in much the same way as that outlined in Figure 3.10. Capillary electro-phoresis separates charged compounds with high resolution and is com-patible with ESI-MS. Problems with coupling capillary electrophoresis with a mass spectrometer stem from the relatively high levels of buffers and salts used that can clog the transfer lines and disrupt the ionisation process. Gel-filled capillaries can be utilised to remove buffers from the ion source. Alternatively, narrow bore (5–10 μm) capillaries are used to reduce the electrolyte flow. The use of low flow rates (nl min^{-1}) in general offers improved sensitivities that are exploited in many applications of electrospray ionisation mass spectrometry.

3.2.11 Low Flow Rate Electrospray Ionisation – Nanospray

Electrospray ionisation mass spectrometry is most often performed using solution flow rates of several μl min^{-1}. There are several advantages, however, in using considerably lower flow rates (10–50 nl min^{-1}). These include compatibility with micro-flow LC and CE separation, improved spray stability, and lower sample consumption. Such experi-ments can also be performed without the use of a sheath gas passing around the spray needle to direct the electrosprayed droplets. Since there is also little solvent to evaporate during the ionisation process, heating the source chamber and/or the use of counter-current dry gases can be avoided.

These so-called *micro-electrospray* or *nanospray* experiments can also be performed off-line by spraying liquids from microfine capillaries prepared by etching or drawing out glass capillaries (Figure 3.11). The capillary is loaded with approximately 1 μl of analyte solution and the flow rate is maintained by the electrospray process without the need for a delivery device such as a syringe pump. The capillary is either coated with a conducting material or a non-corrosive conductive wire is passed through the capillary to supply the high voltage to the tip. A camera or microscope can be used to position the capillary a few millimetres from the entrance lens to the mass analyser.

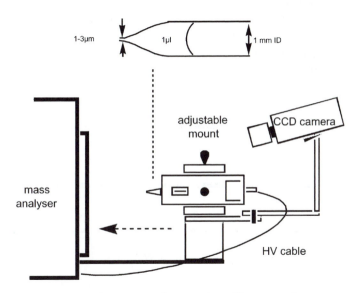

Figure 3.11 *Nanospray needle is mounted on an adjustable support for positioning a few millimetres from the entrance to the mass analyser*

A comparison of the common ionisation techniques is presented in Appendix 3. Once molecules have been introduced into a mass spectrometer as their ions, a mass analyser is used to guide them to the detector through the application of electric and magnetic fields.

3.3 MASS ANALYSERS

Once ions have been formed and introduced into a mass spectrometer, a mass analyser is used to separate them based upon their mass-to-charge ratio through the application of electric and magnetic fields. There are a number of different mass analysers described in the following sub-sections. Many modern instruments feature several mass analysers coupled together for use in tandem mass spectrometry (see Chapter 4) and other applications. When a mass spectrometer is constructed of several mass analysers of a different type, it is referred to as a *hybrid* instrument. This section begins with the simplest mass analyser, the time-of-flight tube.

3.3.1 Time-of-Flight

As the name implies, time-of-flight (TOF) mass spectrometers separate ions and measure their *m/z* based on the time they take to pass ("fly") from the ion source to the detector. The flight tube is usually 1–2 m in length and the basis of the separation makes no use of either electric or magnetic fields. Ions are separated in the *field-free region* of the flight tube before reaching the detector. A simple representation of a TOF mass spectrometer is shown in Figure 3.12.

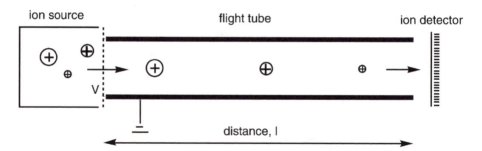

Figure 3.12 *Representation of a time-of-flight (TOF) mass spectrometer*

Ions are first formed in the source and then "pushed" down the flight tube through the application of a high accelerating potential (V) of the same polarity of the ions applied to a lens or grid. The kinetic energy of ions of mass, *m*, and charge, *z*, is given by equation 3.15 where *v* is their velocity. All like-charged ions (common *z*) share the same initial kinetic energy (KE) as they leave the ion source.

$$\text{KE} = 1/2mv^2 = ze\text{V} \qquad (3.15)$$

The time, *t*, is takes for the ions to pass the length of the tube (*l*) is given by $t = l/v$. Substituting for *v* in equation 3.15 and rearranging, leads to equation 3.16.

$$t^2 = m/z \,(l^2/2\,e\,V) \qquad (3.16)$$

Since the length of the flight tube (l) and accelerating voltage (V) are fixed, the time it takes for the ions to reach the detector depends only on their mass and charge. If the time it takes for at least two ions of known mass-to-charge ratio to reach the detector is measured, the time scale can be correlated with m/z values. As a general rule of thumb, singly-charged ions of molecules of 10,000 Da take about 100 μsec to reach the detector. Common accelerating voltages are of the order of 10–30 kV.

As one might expect, a flight tube of a common length (1–2 m) would not be able to separate ions with very similar mass-to-charge ratios. An additional complication arises since, due to the spatial distribution of ions in the ion source and their proximity to the applied electric field, not all the ions receive the same initial kinetic energy. These factors give rise to relatively poor mass resolutions of the order of 100–500 in linear TOF mass spectra. This leads to components in mixtures being unresolved from one another and large errors (~1%) in molecular weight measurements.

To overcome this, several features are now built into most TOF mass analysers that considerably improve mass resolution and thus mass accuracies. The first of these is an *ion mirror, ion reflector* or simply a *reflectron* (Figure 3.13). A reflectron is constructed of a stack of donut-shaped lens connected by a series of resistors across which a high voltage (V_R) is applied. In most instances the voltage difference between each lens of the stack is identical creating a linear or homogeneous field. So-called curved or inhomogenous field reflectrons have also been developed and have some advantages for transmitting ions across a wide m/z range. The potentials applied across the lenses of the reflectron causes the ions that enter it to be gradually repelled. These ions are reflected down the same or second flight tube to a second detector. Improvements in mass resolution are achieved because ions of different kinetic energies

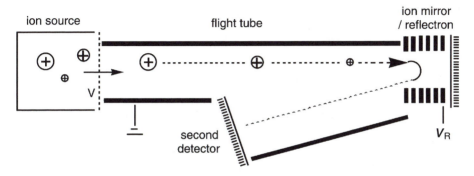

Figure 3.13 *Representation of a reflecting time-of-flight (rTOF) mass spectrometer*

penetrate the mirror to differing degrees. Furthermore, the reflectron effectively extends the flight tube to almost twice its length which (from equation 3.16) can be seen to have a dramatic effect on an ion's flight time. Note that the instrument can also be operated in a linear-mode since, where no voltage is applied to the reflectron, ions pass through it to the first detector.

Consider for the moment two ions of the same *m/z* that have slightly different initial kinetic energies, KE_1 and KE_2 (and velocities v_1 and v_2) when they leave the ion source, where $KE_1 > KE_2$. Ions with greater kinetic energy will pass further into the reflectron before being repelled, while ions with less kinetic energy will travel over a shorter distance. This difference in the flight path and time corrects for the differences in the kinetic energies of the ions so that they reach the detector at the same time.

A dramatic improvement in mass resolution is evident in TOF mass analysers operating in the reflectron over the linear mode as is illustrated for a segment (residues 18–39) of the peptide adrenocorticotropic hormone (ACTH) (Figure 3.14). The best mass resolution is achieved when ions spend equal times in the reflectron and the flight tube of the reflecting time-of-flight (RTOF) analyser.

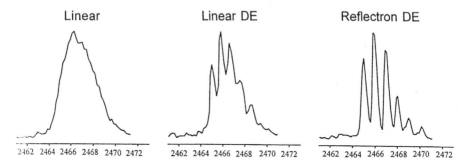

Figure 3.14 *TOF mass spectra of the [M+H]⁺ ions of the peptide ACTH18–39 recorded in the linear (left), linear with delayed extraction (DE) (centre) and reflectron DE mode (right)*

A second feature to improve mass resolution that is employed today in most TOF instruments is *time-lag focusing* (TLF). Time-lag focusing has been revisited in recent years to improve the performance of MALDI-TOF experiments and has been described by different names including *delayed or pulsed-ion extraction* (DE or PIE). In conventional experiments, ions are extracted from the ion source through the application of an accelerating potential immediately after they are formed. If a time delay is introduced before the application of this potential, ions formed with more initial kinetic energy and greater velocities will move further from the ion extraction lens or grid. Application of an accelerating potential pulse imparts more energy into the ions further from the lens than those closer to it. The amplitude is adjusted so that the initially less-energetic ions further from the lens will catch up to the initially more-energetic ions so that they all reach the detector at the same time (Figure 3.15). The improvement in mass resolution that can be attained is illustrated in Figure 3.14.

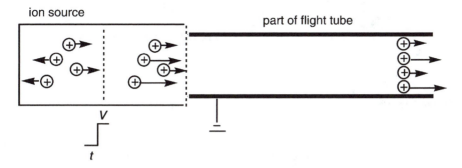

Figure 3.15 *Principle of time-lag focussing (TLF) on a time-of-flight mass spectrometer*

Mass resolutions of up to 30,000 (at FWHM) can now be achieved on TOF mass spectrometers by making use of both time-lag focusing and ion reflectrons. This is a dramatic improvement over their historical mass resolving capabilities. Consequently, TOF instruments can be described as analysers that achieve modest to high mass-resolutions that are second only to those of magnetic-based instruments.

3.3.2 Magnetic Sector

Instruments that contain a magnet positioned over one region of the ions' flight path are the oldest type of mass spectrometer. An ion of charge z moving with a velocity, v, that transverses a magnetic field B at right angles to the direction of the field will experience a centrifugal force given by $zevB$. When this force is equal to the centripetal force, ions adopt a circular path of radius r (equation 3.17).

$$zevB = (mv^2)/r \qquad (3.17)$$

When equation 3.17 is rearranged, equation 3.18 is produced.

$$r = (mv)/zeB \qquad (3.18)$$

This equation indicates that for ions of a particular charge z moving through a fixed magnetic field, B, the radius of their path is dependent only upon their momentum mv. In other words, ions of the same charge will follow a different path when they move pass the magnet influenced only by their mass and velocity (Figure 3.16).

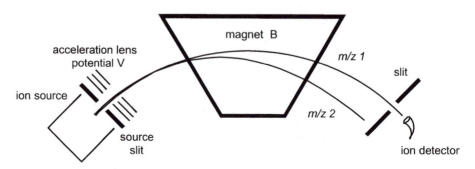

Figure 3.16 *Passage of ions of different m/z through a magnetic field of a single focusing mass spectrometer*

Since the initial kinetic energy of the ions $1/2mv^2$ equals zeV, the initial velocity of the ions is dependent on the potential, V, through which they are accelerated. Rearranging for v^2 we arrive at equation 3.19.

$$v^2 = (2zeV)/m \tag{3.19}$$

From equation 3.17 we can derive:

$$v = (zeBr)/m \tag{3.20}$$

If we square both sides of equation 3.20 we get:

$$v^2 = (zeBr)^2/m^2 \tag{3.21}$$

If equations 3.19 and 3.21 are combined and rearranged, we arrive at equation 3.22.

$$m/z = eB^2r^2/2V \tag{3.22}$$

Therefore specific values of V or B allow ions unique in mass-to-charge to pass through the magnetic field along a path to the detector. Variations in either V or B will cause these same ions to follow a different trajectory and collide with the walls of the flight tube. In Figure 3.16, only ions following the centre trajectory reach the detector at any set of

V and B values. In practice, a series of slits are used throughout the instrument to further improve the focusing and separation of ions.

It follows that a complete mass spectrum, in which all ions in turn are passed to the detector, can be recorded by changing (scanning) V or B over time. In practice, when the accelerating voltage is too low insufficient numbers of ions will leave the ion source and reach the detector. For this, and other reasons, scanning of the magnetic field B is preferred. However, the scan rate of this type of mass analyser is limited by *hysteresis* where the magnetic field can become perturbed. To minimise this, magnetics are scanned more slowly than other mass analysers with time allowed between scans to "settle" the field. Laminated magnetics, however, allow for more rapid scan rates, approaching 0.1 sec decade^{-1} (where a decade is equal to a range covering an order of magnitude difference in mass units, *e.g.* 100 to 1000 u), to be achieved.

As mentioned earlier, ions leave the source with a range of kinetic energies rather than a single value due to their spatial distribution. Since some ions of the same mass will have different velocities and will still reach the detector for a particular set of B and V values, the mass resolution achieved by a single magnet is compromised. To minimise this problem, most modern magnetic sector mass spectrometers also feature an electrostatic or electric sector.

If a radial electrostatic field E is created by two curved plates held at oppositely charged potentials ($+E$ and $-E$), an ion of charge z moving with a velocity v will transverse the field when its electrostatic force equals the centripetal force (equation 3.23).

$$zeE = (mv^2)/r \qquad (3.23)$$

Since the kinetic energy of an ion $1/2mv^2$ equals zeV, equation 3.24 becomes:

$$r = 2V/E \qquad (3.24)$$

Note that the trajectory of the ion defined by r is independent of its mass and charge. At a fixed accelerating voltage, V, the ion's trajectory is thus dependent only on the electric field strength.

Mass spectrometers that combine electric and magnetic sectors are known as *double-focusing* instruments. In most cases, the electric sector is positioned before the magnetic sector in terms of the direction that the ions travel (Figure 3.17). These instruments are referred to as *forward geometry* instruments. Mass spectrometers with the reverse order of sectors are termed *reverse geometry*. The order of mass analysers after

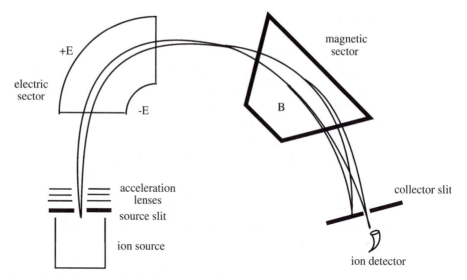

Figure 3.17 *Passage of ions through a double focusing (EB) magnetic sector mass spectrometer*

the ion source is often abbreviated simply as EB or BE. Mass spectrometers can be constructed of even more (three, four, five and even six!) sectors that have particular uses for tandem mass spectrometry (Chapter 4) and studies of ion chemistry (Chapter 6). These are denoted EBE, BEB, BEE, EBEB or BEBE *etc.*

Double-focusing instruments can be scanned in a number of ways but a common scan is one in which the electric and magnetic fields are varied such that the ratio of the field strengths is always held constant (B/E = constant). These scans are known as *linked scans* that will be returned to later in Chapter 4 in the context of tandem mass spectrometry (MS/MS) experiments on magnetic sector mass spectrometers.

Double-focusing sector mass spectrometers can achieve mass resolutions up to 100,000 that allows ions which share the same nominal mass but different exact mass to be resolved. Accurate mass measurements (see Section 5.1) can be obtained where an ion's mass is measured to six decimal places (or a few parts-per-million (ppm)). This can be useful to identify the composition of an ion as discussed in Section 1.3.1.

Reverse geometry sector mass spectrometers are also useful in mass-analysed ion kinetic energy spectra (MIKES) experiments in which metastable decomposition products are detected (see Section 4.3.1).

3.3.3 Quadrupoles

The concept of the quadrupole mass analyser was first reported by Paul and Steinwedel in the 1950s. A quadrupole (Q) consists of four rods arranged in parallel where those opposite to one another are electrically connected (Figure 3.18). The quadrupole has a number of advantages over magnetic sector mass analysers including the low cost of construction, their compact size, and fast scanning capability.

A voltage of opposite polarity ($+/-V$) is applied to adjacent rods consisting of a direct current (DC) component (denoted U) and a radio-frequency (RF) (denoted $V_{RF}\cos(wt)$) component where w is the angular frequency of the RF field. Ions are accelerated out of the ion source along the z-axis between the rods. They experience forces in the x and y direction $-ze(\mathrm{d}V/\mathrm{d}x)$ and $-ze(\mathrm{d}V/\mathrm{d}y)$ that cause them to oscillate toward and away from the rods. When the oscillation becomes too large the ions strike the rods and do not reach the detector.

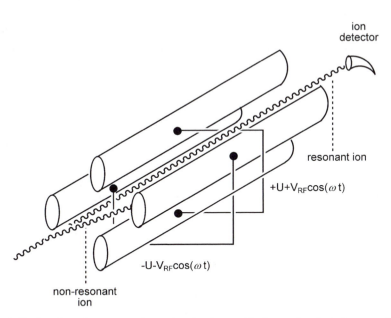

Figure 3.18 *Quadrupole mass analyser showing ion oscillation under the influence of the variable fields*

For any given analyser, the radius of an imaginary cylinder that fits in the centre of the rod (r_0) is constant as is the frequency of the RF field w. Two functions a and q define a stable trajectory for which ions do not collide with the rods across a range of values for U and V_{RF} (equation 3.25 and 3.26).

$$a_z = -2a_r = -4zeU/m^2r_0^2w^2 \qquad (3.25)$$

$$q_z = -2q_r = -2zeV_{RF}/m^2r_0^2w^2 \qquad (3.26)$$

A plot of a versus q is known as a stability diagram where the regions below the curves for m_1 m_2 and m_3 represent the values of a and q for which ions follow a stable ion trajectory to the detector (Figure 3.19). In principle, the quadrupole can be operated over a range of values of U and V_{RF} such that the (a,q) coordinates are always below the curves. In practice, the voltages U and V_{RF} are held at a fixed ratio to maximise the mass resolution that can be achieved. This leads to an operating region defined by a line with a slope of $2U/V_{RF}$. A complete mass spectrum is obtained by scanning the voltages U and V_{RF} where this fixed ratio is maintained. Ideally the voltages U and V_{RF} should be held constant throughout the passage of ions of a particular m/z through the analyser. Scan rates of the order of 1000 u sec^{-1} are common and allow for the majority of ions of a particular m/z to reach the detector.

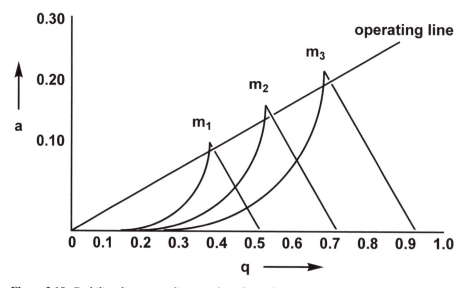

Figure 3.19 *Stability diagram indicating the values of a and q for which ions of mass m_1, m_2 and m_3 follow a stable trajectory to the ion detector*

Because quadrupoles operate at lower voltages and can be scanned at faster rates than magnet-based mass spectrometers, they are more easily coupled to gas and liquid chromatography instruments. However, they achieve lower mass resolutions of up to approximately 5000.

3.3.4 Quadrupole Ion Trap

There are strictly two types of ion traps though the term is usually associated with only one of them: the *quadrupole ion trap*. Quadrupole ion traps (QIT or IT) are so named because they use similar operating principles to those of the standard quadrupole mass analyser. Despite this they are constructed very differently and consist of two conical lens or electrodes, and one "donut-shaped" ring lens (Figure 3.20). Ion cyclotron resonance (ICR) mass spectrometers are also ion traps but use magnetic fields to store ions as described in the next section.

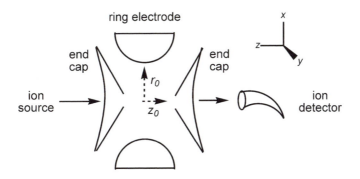

Figure 3.20 *Cross-section of a quadrupole ion trap*

In a QIT, ions are held or trapped in the small interior volume between the conical lenses that form the "caps" of the trap and the centre of the ring electrode. By lowering and raising the voltages on the entrance and exit trap electrode, ions can pass into the trap, be stored for some period of time (usually μs), and then be released to the detector. The trap is usually operated in the *mass selective stability mode* where ions of a particular *m/z* are selectively released from the trap.

Within the trap ions undergo a complex sinusoidal motion with the application of an oscillating RF potential to the ring electrode. An ion will be stored in the trap depending upon the values for the mass, *m*, and charge, *z*, of the ion, the radius of the ring electrode (r_0), the separation of the caps from the centre of the ion trap (z_0), the oscillating frequency, *w*, the amplitude of the potential applied to the end caps *U*, and the amplitude of the ring electrode voltage, V_{RF}, according to equations 3.27 and 3.28.

$$a_z = -2a_r = -16zeU/m(r_0^2 + z_0^2)w^2 \tag{3.27}$$

$$q_z = -2q_r = 8zeV_{RF}/m(r_0^2 + z_0^2)w^2 \tag{3.28}$$

Again an ion stability diagram can be constructed to define the (a,q) coordinates for which ions are stored in the trap. Ions possessing values of a and q that give them both axial (along the z-axis between the caps) and radial (in the plane of the ring electrode) stability will remain trapped. In modern ion trap instruments, an inert gas such as helium (added to a pressure of about 10^{-3} Torr, or 0.1 Pa) helps to store the ions by lowering their kinetic energy through collision with the gas.

A unique feature of traps is that ions can be introduced and stored until a sufficient density is achieved for ejection and analysis. In conventional quadrupole and magnetic sector instruments, ions continuously pass through the mass spectrometer and their population is largely determined by the efficiency of the ionisation process. The ability to store ions, and the relatively short path to the detector that the ions follow in a quadrupole ion trap, can lead to very high detection sensitivities. Ion traps can also be exploited in studies of ion chemistry (see Chapter 6) by using longer containment times.

In practice, it is necessary to balance the desire to store large numbers of ions in the trap with *space charge effects* that arise from the repulsion of neighbouring ions. These repulsive forces cause ions to leave the trap and not be detected. It also perturbs the motion of ions in the trap resulting in degraded mass resolution and mass shifts that lead to large errors in mass measurements. Thus it is desirable to control the population of ions in the trap at all times during analysis. For this reason, *automatic gain control* (AGC) was developed to control the ion generation rate and couple it to the time period in which ions are introduced into the trap and stored. A typical operation of an ion trap involves lowering the end cap potential closest to the ion source to allow ions to enter the trap. This voltage is raised once sufficient numbers of ions are stored as determined by the AGC measurements. The ring electrode is held at an appropriate value to store ions over a range of m/z values. The potential on the second cap electrode is then lowered and the ring electrode voltage V_{RF} is linearly ramped to eject ions of increasing m/z in turn onto the detector.

3.3.5 Ion Cyclotron Resonance

Traps that make use of a magnetic field are known as *ion cyclotron reson-ance* (ICR) mass analysers. The name derives from the frequency of an ion's circular motion w_C within a magnetic field, B. In these instruments the trap consists of a cubic, rectangular or cylindrical "box" with entrance and exit slits cut in opposite sides or plates (Figure 3.21).

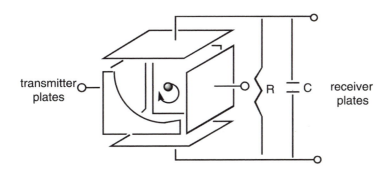

Figure 3.21 *Ion trap of an ion cyclotron resonance (ICR) mass spectrometer*

In a magnetic field, B, an ion with a velocity v will adopt a circular trajectory with a radius r perpendicular to the field when the centripetal and centrifugal forces it experiences are equal (equation 3.29).

$$zvB = (mv^2)/r \tag{3.29}$$

The angular velocity of the ion perpendicular to the field is given by equation 3.30.

$$w_C = v/r \tag{3.30}$$

Substituting equation 3.30 into equation 3.29 and simplifying, equation 3.31 is derived.

$$w_C = zB/m \tag{3.31}$$

Thus an ion's cyclotron frequency depends on its mass and charge but is independent of its velocity (equation 3.31) and the m/z ratio for an ion can be determined by measuring its cyclotron frequency. Ions of lower m/z have higher cyclotron frequencies (Figure 3.22A); ions with higher m/z have low cyclotron frequencies (Figure 3.22B). Note that oppositely charged ions would move with the same cyclotron frequency but in opposite directions were they both present in the trap.

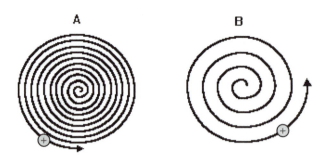

Figure 3.22 *Ion cyclotrons of high frequency or low m/z (A) and low frequency or high m/z (B)*

Ions moving in a magnetic field adopt stable cyclotron orbits and can be stored (up to several hours!) provided the pressure is kept low (typically $<10^{-8}$ Torr or 10^{-6} Pa). They do not, however, generate any detectable signal. In order to generate an electrical signal or current, ions of a particular *m/z* ratio are given extra energy through the application of an oscillating electrical field, E_t, (equation 3.32).

$$E_t = E_0 \cos w_C t \qquad\qquad (3.32)$$

If, and only if, the frequency of field is the same as their cyclotron frequency, the ions absorb energy increasing their velocity and orbital radius while maintaining a constant cyclotron frequency. As these ions approach the top plate of the cell, electrons are attracted to the plate from ground. When the ions circulate towards the bottom plate, the electrons travel back down to the bottom plate. This coherent motion of ions between the two plates produces an electrical current that can be amplified and detected and is known as the *image current*. The amplitude of the current is proportional to the number of ions in the cell at this frequency. This phenomenon provides the basis for ion cyclotron resonance mass spectrometry with ions of different cyclotron frequencies unaffected.

The excitation of ions by an oscillating electric field has three main objectives in ICR mass spectrometry. First it accelerates the ions coherently to a larger orbital radius in the trap so they can be detected. Second the kinetic energy of the ions is increased to achieve their dissociation or to facilitate ion molecule or ion-ion reactions where desirable. Thirdly, it allows ions to be accelerated to radii larger than the trap to effect their removal from or collision with the trap.

In an ideal vacuum, excited ions would maintain their orbiting trajectories indefinitely. In practice, their energy is dampened and they

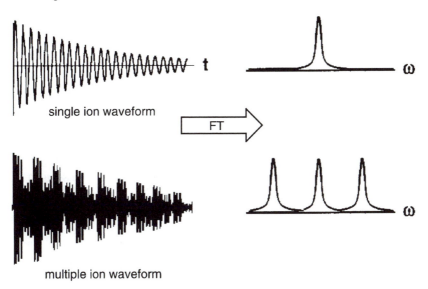

single ion waveform

FT

multiple ion waveform

Figure 3.23 *Single and multiple ion waveforms and frequency domain spectra after application of a Fourier transform*

return to their original energy and cyclotron frequency. Therefore a plot of the amplitude of the frequency over time resembles that shown in Figure 3.23.

A complete mass spectrum is derived from the overlap of the individual ion waveforms shown in Figure 3.23. Such complicated data are best devolved through the application of a mathematical process known as a Fourier transform (FT). In a Fourier transform ion cyclotron resonance mass spectrometer, all ions in the trap are excited simultaneously to larger orbits by applying a wide range of excitation energies at once (a *broadband excitation*) in the form of an excitation *chirp*. The excitation event is very brief so that ions are not excited to orbits greater than the dimensions of the cell. Applying the Fourier transform, the combined waveform can be devolved to a frequency domain (w) spectrum from which a mass spectrum (m/z) can be assembled. As a result FT-ICR mass spectrometry (also known less descriptively as FT-MS) can detect and measures the m/z ratios of all ions in the trap simultaneously; a unique feature of this particular mass analyser.

The mass resolution ($m/\Delta m$) achieved in this type of mass analyser can be defined by equation 3.33 where Δw is the width of the peaks in the frequency domain spectrum.

$$m/\Delta m = zB/m\Delta w \qquad (3.33)$$

Note that mass resolution increases proportionately with an increase in the magnetic field strength, *B*. Other performance improvements are attained by using a high magnetic field that include an increase in the number of ions that can be trapped, an increase in the maximum trapping time that is possible, and an increase in the speed of data acquistion. Therefore many modern FT-ICR mass spectrometers feature magnets that operate at 9 or even 14 Tesla. An impressive demonstration of the advanced capabilities of an FT-ICR mass spectrometer is the unit resolution of the isotope peaks for ions of large proteins, including elucidation of the protein's elemental composition after its expression in heavy isotope ^{13}C and ^{15}N-depleted media (Figure 3.24).

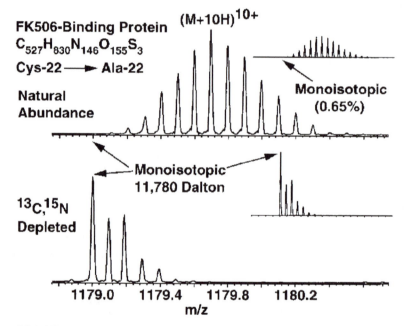

Figure 3.24 *ESI FT-ICR mass spectrum (at 9.4 Tesla) of a mutant (C22A) FK506-binding protein. Top: Natural-abundance isotopic distribution. Bottom: Isotopic distribution for the same protein grown in a medium of ^{12}C (99.95%) and ^{14}N (99.99%) together with that predicted based on chemical formula (inserts)*
(Source: A.G. Marshall, M.W. Senko, W. Li, M. Li and S. Dillon, Protein molecular weight to 1 Da by 13C, 15N double-depletion and FT-ICR mass spectrometry, *J. Am. Chem. Soc.*, 1997, **119**, 433–434)

3.3.6 Hybrid Instruments

It is often advantageous to construct mass spectrometers that consist of more than one mass analyser. Where the individual mass analysers are of a different type (TOF, magnetic sector, quadrupole, ion trap or ICR) the combined mass spectrometer is known as a *hybrid* instrument. Since different mass analysers offer complementary features in terms of mass resolving power, mass range and ion transmission, a combination of different mass analysers in a single instrument can lead to performance enhancements and/or benefits for certain experiments, particularly tandem mass spectrometry described in Chapter 4.

In principle, any combination of two or more *different* mass analysers produces a hybrid instrument but in practice some are more suitable for coupling than others. A few hybrid instruments in common use today are those featuring combinations of magnetic sectors and quadrupoles (such as the configuration BE-Q, consisting of a magnet, electric sector and quadrupole), and a quadrupole and time-of-flight analyser (Q-TOF).

A comparison of the performances of the various mass analysers is represented in Appendix 4. Now that the formation and separation of ions within a mass spectrometer has been considered, it is time to understand in some detail how they are detected.

3.4 DETECTORS

Once ions leave the mass analyser they pass to the ion detector (except in the case of the ICR) where they generate an image current. The earliest mass spectrometers detected ions by means of photographic paper onto which line images were visualised. Many different types of detectors are now in use, the choice of which depends in part on the nature of the mass analyser. To begin, the simplest detector used in mass spectrometry, the so-called *Faraday cup or cage* will be considered.

3.4.1 Faraday Cup

The Faraday cup or cage consists of a long thin rectangular box arranged such that the incoming ion beam strikes the base of the box at the *collector*. The collector plate is usually angled steeply with respect to the ion beam in order to prevent reflected ions and secondary electrons emitted from the plate from escaping. Each collector box is typically enclosed by a second box (held at ground) that serves to shield the detector from electrical noise. A slot on the front of this box acts as a resolving slit. When positive ions strike the collector plate they are neutralised by electrons drawn from ground across a resistor. This current is amplified and recorded. Faraday cup detectors are extremely simple in their design and are an inexpensive and reliable alternative to other detectors. They are best used for quantitation and accurate mass measurements where ion currents do not change appreciably during the course of a measurement. The smallest ion current that can be detected is of the order of 10^{-15} A. Faraday cups are less suited to experiments where the ion current changes quickly, as is the case in experiments conducted with online chromatography where the mass analyser is rapidly scanned and the population of ions transmitted to the detector varies.

3.4.2 Electron Multipliers

Electron multipliers (EM) achieve higher detection efficiencies than Faraday cups by amplifying the secondary electrons emitted from the detector surface. There are two common types; one in which the detector surface is constructed of a series of discrete plates (*discrete dynode*) connected by chain of resistors, and a second in which a single, continuous detector surface is used (*continuous dynode*) (Figure 3.25).

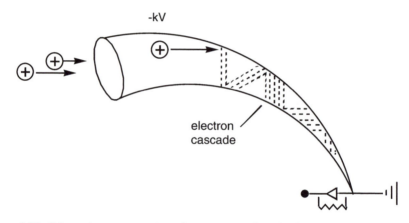

Figure 3.25 *Schematic representation of a continuous dynode electron multiplier*

In the discrete dynode detector, a high voltage applied across the resistor chain leads to an equal voltage difference between each successive plate surface. The plates are constructed principally of a beryllium/copper alloy or aluminium onto which oxides of beryllium or aluminium coat the surface. The multiplier must not be exposed to air when operational as this can damage the coatings and prevent its further use. The multiplier is assembled such that the dynodes have progressively higher voltages from the first to the last (over some 10–20 plates). As ions strike the first plate, secondary electrons are emitted from the surface and projected to the second dynode plate. This process repeats itself leading to a *cascading effect*. The amplified current (with gains of typically 10^7 to 10^8 during the detector's useful life of operation) is finally detected. In the alternate continuous dynode electron multiplier, the internal surface acts as a resistor so that an electrical gradient is achieved across the detector. The secondary electrons emitted from the collision of ions with the detector surface are accelerated to the tip of the detector amplifying the current.

Current gains of the order of 10^5 are typically achieved after which the detector response plateaus. Electron multipliers are commonly used in conjunction with scanning quadrupole and magnetic sector mass analysers. A disadvantage of these detectors is that the emission of secondary electrons is dependent on the mass and charge of the incident ions. That is, ions with a *m/z* ratio of 500 will produce a different detector response than ions with a *m/z* of 5000. This non-linear detector response leads to difficulties in quantitation experiments, where the ion signal is used to reflect the levels of each component in a sample mixture.

3.4.3 Microchannel Plate Electron Multipliers

Microchannel plate (MCP) detectors consist of a thin flat plate containing series of small (~10 μm) channels (Figure 3.26). The plates that differentiate the channels are typically angled at several degrees to the incident ion beam. Ions pass into any of the channels, strike the plate walls and emit electrons that are deflected to the opposite plate. This process repeats itself thereby amplifying the electron current. Gains of the order of 10^5 can be achieved with one plate. Amplification can reach 10^8 using a stack of several plates.

These devices are commonly used in TOF mass analysers since ions are unfocused and arrive at the detector over large areas. They also are implemented in array detectors.

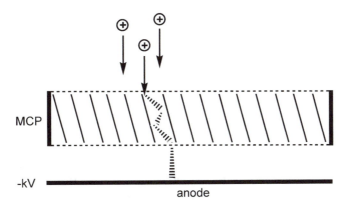

Figure 3.26 *Schematic representation of a micro-channel plate (MCP) detector*

3.4.5 Array Detectors

A limitation of scanning mass analysers (quadrupoles and magnetic sectors) over non-scanning TOF instruments, is that only ions of a particular m/z ratio are detected at any particular point in time. This compromises sensitivity since many ions passing into the mass spectrometer are actually deflected from the detector and go undetected.

Beginning in the mid 1970s, array detectors were constructed for magnetic sector mass spectrometers that enabled all ions, or ions over a large range of m/z ratios, to be detected simultaneously.

Photodiode arrays (PDA) are the most common form in which a microchannel plate detector is coupled to fibre optical channels *via* a phosphorescent screen (Figure 3.27).

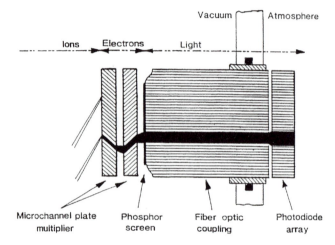

Figure 3.27 *Schematic representation of a photodiode array (PDA) detector*
(Source: S. Evans, in *Methods Enzymology*, McCloskey (ed), Academic Press, New York, 1990, Vol. 193, Ch. 3, p. 80, Figure 9)

Ions enter the microchannel plates and the electrons emitted strike the phosphorescent screen. The screen consists of a layer of aluminium coated with crystalline phosphor often composed of caesium iodide (CsI) and thallium (Tl). This surface emits light in the form of photons that are transmitted down the fiber optic channels onto a charge or plasma-coupled device (CCD or PCD). The photodiodes of the CCD or PCD, unlike the phosphor screen of the image intensifier, react independently of each other. Each pixel of the photodiode array is typically 25 µm by 2.5 mm in size and converts the photons to charge. The photodiodes of the CCD array record the charge necessary to neutralise that accumulated in each pixel and this data is integrated and stored.

Array detectors constructed for magnetic sector mass spectrometers are able to detect ions across some 5% of the full m/z range of the instrument. Detection ranges of approximately 25% are possible for quadrupole instruments albeit with some mass discrimination; that is, the preferential detection of ions at one end of the range over the other.

Other array detectors, such as the position and time-resolved ion collector (PATRIC) have been developed for commercial use. Like the photodiode array, ions first strike a microchannel plate. The electrons emitted from the MCP are accelerated to a collector stack consisting of a series (some 50) of conductive strips connected together by a capacitor chain. Charge amplifiers located at each end of the stack detect the current passing through the capacitors. The time the ions take to pass through the array depends on their mass and charge so that ions of a particular m/z value can be tracked as they pass across the array.

The performance of this array is similar to a PDA. All array detectors result in some compromise in mass resolution since even a focused ion beam can pass through the array at multiple positions or channels. Unit resolution across most of the mass range can usually be obtained, although at some expense in terms of sensitivity.

3.5 COMPUTER ACQUISITION OF DATA

3.5.1 Role of Computers in Mass Spectrometry

Computers now play a central role in all mass spectrometry experiments. Computers are used for instrument control, data acquisition, data processing and storage. Once the sample is introduced into the mass spectrometer, all remaining functions are performed under computer-control. This includes mass calibration and tuning of the instrument to optimise the ion signal at the detector.

Both personal computers and UNIX-based workstations are used to control mass spectrometers. Mass spectrometry experiments are pre-programmed using the computer software where method files control instrument parameters including source temperatures, gas pressures, lens and detector voltages. The same computer is typically also used to control peripheral devices such as gas and high pressure liquid chromatographs when these are coupled to the mass spectrometer. Computers allow the chromatographic conditions to be pre-programmed and the user to monitor the results at a second detector (such as a UV detector) in addition to the mass spectral data within a common interface.

Computers also allow a multitude of different scan experiments to be performed during a single acquisition. Tandem mass spectra (see Chapter 4) can be recorded where sufficient ion current for a nominated *m/z* is detected in a preliminary MS scan. This enables both molecular weight and structural information to be derived in an automated manner.

3.5.2 Analog-to-Digital Converters

Since the mass spectrometer acquires data in an analog format, an *analog-to-digital converter* (ADC) is used to convert the detector response into a digital format. The ion current read by the detector is amplified and filtered to remove high frequency noise. This current, recorded over the time period of the experiment, is plotted on a mass-to-charge scale by comparison with data obtained from an earlier mass calibration experiment. In a mass calibration experiment, ion signals produced from a well-characterised sample or sample mixture are assigned to their theoretical mass values. This is often performed under computer control with reference to a table of stored theoretical masses.

3.5.3 Data Processing and Interpretation Algorithms

Data interpretation and processing are also significantly aided by the use of computers. Sections of a mass spectrum can be replotted, normalised (where the largest peak is plotted to the 100% level on the *y*-axis), and "smoothed". Smoothing algorithms are applied to filter extraneous noise that appears in a mass spectrum due to electronic interference and background ion current from ions associated with the analyte. A subtraction algorithm can also be applied to remove such signals, where mass spectra representative of the background are pre-recorded independent of the sample ions.

Where the ion current is monitored during multiple scans of the mass analyser over the course of the experiment, an ion chromatogram is also recorded. Processing of this data set enables a mass spectrum to be generated from any one scan or the sum of several scans. A selected ion chromatogram can also be generated from the data set by extracting all time points along the chromatogram during which ions of a particular *m/z* were detected. This is useful when a sample is first subjected to chromatographic or electrophoretic separation prior to mass spectrometric analysis. Components of a sample mixture will subsequently only pass into the mass spectrometer over a limited period of time subject to the performance of the experiment. In other words, a component that elutes early from a chromatographic or electrophoretic column will appear in mass spectra recorded early in the experiment, but not in spectra recorded later. The converse is also true. Thus the time at which a component is introduced and detected within the mass spectrometer can be obtained from the ion chromatogram.

In addition to data acquisition, analysis and processing, many other programs have been written to assist with data interpretation. These include those that deduce the composition of the ion based upon its isotopic distribution, the conversion of *m/z* ratios for multiply-charged ions into a molecular weight value, the structure of an analyte based on tandem mass spectral data (see Chapter 4), and identification of an analyte by comparison of a mass spectrum with data stored in a library or database. The importance of mass spectral libraries and the use of such databases will be discussed in later chapters (Chapters 5 and 8).

3.6 VACUUM PUMPS

An important component of all mass spectrometers, not often considered during operation, are the vacuum pumps. Many instruments use at least two pumps; one to evacuate the instrument chamber of air after assembly or venting and a second to reduce the pressure to that required to operate. Pressures vary within different mass spectrometers, and within each component, but it is typical for instruments to operate over a range of 10^{-3} to 10^{-6} Pa (1 Pa = 133.3226 torr or millimetres of mercury, mmHg). As described in Section 3.1, mass spectrometers operate under vacuum to prevent the collision and reaction of ions with residual gas molecules during their flight from the ion source to the detector. Low pressures also prevent condensation from building up inside the instrument that would coat critical lens and surfaces and create electrical discharges. The detector, too, must be protected from moisture and oxidation for it to operate properly.

3.6.1 Rotary Pumps

The most efficient way in which to reduce the pressure in a mass spectrometer to the 1 Pa level is with a *rotary* or *roughing pump*. In this pump, a rotating inner barrel draws the gas in from one side of a chamber and compresses it on the other (Figure 3.28). The compressed gas passes through a hydrocarbon oil reservoir, where the vapour pressure of the oil

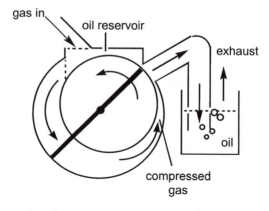

Figure 3.28 *Cross-section of a rotary vacuum pump*

influences the lowest pressure that can be obtained. Rotary pumps have the capacity to dispel approximately 100 litres of gas per minute through the pump.

3.6.2 Diffusion Pumps

Diffusion pumps are used in conjunction with a roughing or rotary pump and operate at pressures below 1 Pa. Unlike rotary pumps, diffusion pumps use no moving parts and operate by expelling the vapour trapped in a high-boiling oil reservoir. Vapours pass up a chimney, cool and

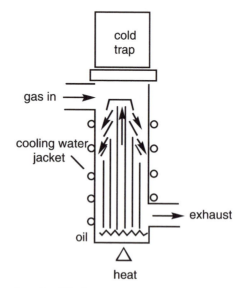

Figure 3.29 *Cross-section of a diffusion pump*

saturate the gas molecules in the pump dragging them to the base. This compresses the gas in the pump. The oil vapour is continually recycled to achieve pressures of the order of 10^{-6} Pa (Figure 3.29).

A serious problem with diffusion pumps is that should the instrument become vented while in operation, oil would be dragged from the reservoir into the spectrometer. A cold trap is generally positioned at the head of the pump to avoid oil and vapour from entering the mass spectrometer. Preventing the oil from reaching a high temperature is also necessary to avoid problems. This is achieved by placing a water-cooled jacket around the pump and through the application of synthetic polyphenyl and silicone lubricants that can better withstand continual heating and cooling.

3.6.3 Turbomolecular Pumps

A popular high vacuum pump is a *turbomolecular pump* often known simply as a *turbo*. Turbos are constructed of a turbine compressor, like that used in a jet engine, and are often preferred over a diffusion pump for several reasons. First, the pumps use no oil exposed to vacuum. Second, the pump can attain full operation more quickly than a diffusion pump leading to faster evacuation times whenever a mass spectrometer is opened for repair or service.

Within a turbo pump, a fast moving rotor or set of rotors (operating at typically 50,000 revolutions per minute) draws gas through the pump at a capacity of the order of $10-100\,l\,s^{-1}$ (Figure 3.30). Because of the speed of the rotors, the pump must be machined to low tolerances in order for it to operate efficiently for several years. The gas is further passed

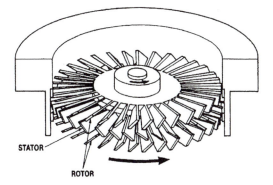

STATOR

ROTOR

Figure 3.30 *View of the rotor of a turbomolecular pump*
(Source: J.T. Watson, *Introduction to Mass Spectrometry*, 3rd edition, Lippincott-Raven, Philadelphia, 1997, Figure 19.7, p. 431)

through a roughing pump to the exhaust. It is important that the pump is vented to atmospheric pressure before it is stopped to prevent oil being dragged from the rotary backing pump into the turbo. Pressures of the order of 10^{-6} to 10^{-7} Pa can be obtained. Turbomolecular pumps are commonly used in quadrupole and magnetic sector mass spectrometers. In the case of the latter, multiple turbo pumps flank regions of the flight tube to maintain suitable operating pressures throughout.

3.6.4 Cryopumps

Among the pumps that offer the lowest pressure are *cryogenic pumps*. Cryogenic pumps achieve pressures as low as 10^{-9} Pa by condensing residual gases on surfaces maintained at extremely low temperature. These coldheads operate at about 20 K (−278 °C) by means of liquid helium. The low temperatures are maintained by expanding the helium gas held at high pressure. As helium passes through heat exchangers (made of a fine metal mesh with a high heat capacity) on its way back to the refrigeration compressor, heat is removed thereby cooling the gas. A typical design construction is shown in Figure 3.31.

Cryopumps create a vacuum by condensing and freezing most of the gases in the vacuum chamber on several helium-cooled coldheads. The

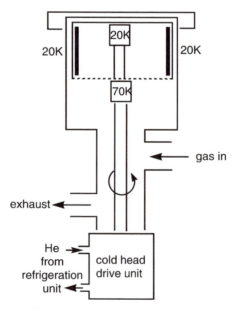

Figure 3.31 *Cross-section of a cryopump*

process used is the same familiar process that causes water vapour to condense on a bathroom mirror.

The cryopump condenses water vapour in the pump's first stage. The first stage operates at a temperature of 70 K (−228 °C). Most gases in air (O_2, N_2, *etc.*) in the vacuum chamber are frozen onto a second-stage condensing array by an identical process. The second stage operates at a temperature of around 20 K (−278 °C). Since the gases hydrogen, helium, and neon cannot be frozen at these temperatures a portion of the second stage contains a surface coated with highly porous activated charcoal. These gases are absorbed onto the charcoal and removed from the vacuum chamber. As the charcoal becomes saturated with gas, however, the pump's ability to efficiently maintain low pressures begins to deteriorate. To overcome this, after operation for a day or two, the cryopumps of a mass spectrometer are vented and the coldheads warmed to release the condensed and absorbed gases.

Cryopumps are commonly used today in FT-ICR mass spectrometers where very low pressures are required during ion detection periods to minimise ion-molecule and ion-ion interactions that can dampen an ion's coherent motion. This allows the characteristic high mass resolutions of these analysers to be achieved.

FURTHER READING

J.T. Watson, *Introduction to Mass Spectrometry*, 3rd edition, Lippincott-Raven, Philadelphia, 1997

E. DeHoffmann and J. Charette, *Mass Spectrometry – Principles and Applications*, 2nd edition, John Wiley & Sons, New York, 2001.

A.G. Harrison, *Chemical Ionization Mass Spectrometry*, 2nd edition, CRC Press, Boca Raton, Florida, 1992.

H.D. Beckey, *Principles of Field Ionization and Field Desorption Ionization in Mass Spectrometry*, Pergamon Press, Oxford, 1977.

R.D. Macfarlane and D.F. Torgerson, Californium-252-plasma desorption time-of-flight mass spectrometry, *Int. J. Mass Spectrom. Ion Phys.* 1976, **21(1–2)**, 81–92.

M. Barber, R.S. Bordoli, R.D. Sedgwick and A.N. Tyler, Fast atom bombardment of solids (FAB): a new ion source for mass spectrometry, *J. Chem. Soc. Chem. Commun.*, 1981, 325–327.

K. Tanaka, H. Waki, Y. Ido, S. Akita, Y. Yoshida and T. Yoshida, Protein and polymer analyses up to m/z 100,000 by laser ionization time-of-flight mass spectrometry, *Rapid Comm. Mass Spectrom.*, 1988, **2**, 151.

M. Karas and F. Hillenkamp, Laser desorption ionization of proteins with molecular masses exceeding 10,000 daltons, *Anal. Chem.*, 1988, **60(20)**, 2299–2301.

C.R. Blakley, M.L. Vestal, Thermospray interface for liquid chromatography/mass spectrometry, *Anal. Chem.*, 1983, **55(4)**, 750–754.

M. Yamashita and J.B. Fenn, Electrospray ion source. Another variation on the free-jet theme, *J. Phys. Chem.*, 1984, **88(20)**, 4451–4459.

R. Cotter *Time-of-Flight Mass Spectrometry. Instrumentation and Applications in Biological Research*, American Chemical Society, 1997.

W. Paul, H. Steinwedel, A new mass spectrometer without magnetic field, *Z. Naturforsch.*, 1953, **8A**, 448–450.

R.E. March, R.J. Hughes, *Quadrupole Storage Mass Spectrometry*, Wiley-Interscience, 1989.

R.E. March and J.F.J. Todd (eds) *Practical Aspects of Ion Trap Mass Spectrometry*, CRC Press, 1996.

M. Oehme, *Practical Introduction to GC-MS Analysis with Quadrupoles*, John Wiley & Sons, 1999.

A.G. Marshall, C.L. Hendrickson and G.S. Jackson, Fourier transform ion cyclotron resonance mass spectrometry: A primer, *Mass Spectrom. Rev.*, 1998, **17**, 1–35.

A.G. Marshall and F.R. Verdun, *Fourier Transforms in NMR, Optical, and Mass Spectrometry: A User's Handbook*, Elsevier, Amsterdam, 1990.

J.R. Chapman, *Computers in Mass Spectrometry*, Academic Press, New York, 1978.

CHAPTER 4

Tandem Mass Spectrometry

4.1 BASIC PRINCIPLES; PRECURSOR AND FRAGMENT IONS

With the exception of electron ionisation, all of the ionisation methods described in Section 3.2 lead to the formation of molecular ions (or pseudo-molecular ions) with little fragmentation. This is advantageous from the perspective of measuring the molecular weight of a compound, but a disadvantage if one wants to obtain structural information.

Tandem mass spectrometry was developed to derive structural detail originally through the use of two mass analysers between the ion source and the detector. A region between the two mass analysers is used to effect the dissociation of incoming ions. A schematic representation of a tandem mass spectrometry experiment is shown in Figure 4.1. The first mass analyser is used to transmit only ions of a particular m/z ratio into the dissociation region at any point in time. Within the dissociation region or chamber, ions are "excited" energetically through a number of different processes described within the next section. This leads to the production of fragments through bond cleavage of the precursor ions. The second mass analyser is scanned to pass, in turn, the products of the dissociation onto the detector. Since two mass analysis steps are involved, tandem mass spectrometry is often referred to as an MS/MS experiment and the resulting fragment or product ion spectra are known as MS/MS (or MS^2) spectra.

In practice the components of a sample are first surveyed in the MS-mode, usually by scanning the first mass analyser (MS-1) and passing all ions through MS-2 onto the detector. A series of tandem mass spectrometry experiments are then performed to detect the *fragment* (or *product*) *ions* (historically, *daughter ions*) for each of the *precursor* (or *parent*) *ions* formed in the ion source. The term *fragment ion* is preferred over *product ion* to differentiate them from the products of an ion-molecule reaction that have higher masses. The products of dissociation, in contrast, always have a lower mass since they represent a fragment or part of an intact molecular species.

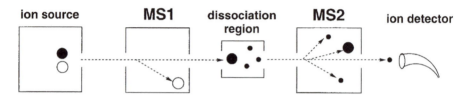

Figure 4.1 *Schematic representation of a tandem mass spectrometry experiment*

To derive structural information from MS/MS spectra one usually has to know some general information about the nature of the compound. Beyond this, the vast numbers of tandem mass spectra that have been obtained and interpreted for known compounds enable dissociation pathways and products to be assigned for an unknown compound with significant reliability. Studies over several decades have shown that certain bonds are more susceptible to dissociation than others, and particular structural features in a molecule can drive the fragmentation of ions in a predictable manner. Where bond cleavage takes place in the vicinity of a charged group, the fragmentation is termed *charge localised*. This is not to say that all molecular ions have charges fixed or localised at only one position; rather the molecular ions formed in the source may represent a mixture in which charge is positioned at different locations within the complete set of ions formed. In contrast, when bond dissociation takes place far removed in the molecule from a charge-bearing site the fragmentation is referred to as *charge remote*. The energy gained during the dissociation process may also lead to some migration of charge in an ion prior to bond cleavage, so some care has to be taken in defining a fragmentation process as charge remote.

These issues will be returned to later in Chapter 7 in tandem mass spectrometry studies of peptides, but first the ways in which an ion can be dissociated on its journey to the detector are considered.

4.2 DISSOCIATION PROCESSES AND THEORY

There are a number of dissociation methods utilised to add energy to ions and cause them to dissociate by bond cleavage. By far the most common of these is collisional activation.

4.2.1 Collisional Activation (CA)

If an ion collides with a neutral atom or molecule, some of the ion's kinetic energy is lost as its translational velocity decreases. The energy lost is converted into internal energy causing the ion to fragment. In the *collisional activation* process, ions are collided with gas molecules (N) held at a moderately high pressure in a chamber or region between the mass analysers. Depending on the velocity of the ions and the density of the gas in the chamber, ions will undergo a few, or multiple, collisions with the gas. The overall dissociation process occurs in two steps, illustrated for positive-charged ions in equation 4.1.

$$m_p^+ + N \rightarrow m_p^{+*} \rightarrow m_f^+ + m_n \qquad (4.1)$$

The first step involves energetic excitation of the precursor ion (m_p^+) through both electronic and vibrational processes, and the second involves the dissociation of the energetically-excited precursors (m_p^{+*}) to a fragment ion (m_f^+) and neutral portion (m_n). In the strictest sense, the first step is known as collisional activation (CA) and the second is referred to as *collisionally-activated dissociation* (CAD) or *collision-induced dissociation* (CID). It is necessary in a CAD experiment to maintain conditions such that the ions are fragmented into a few discernible species rather than obliterated into individual atoms. The latter would serve little use in deciphering the structure of the analyte.

The choice of gas is important in order to prevent the reaction of ions with gas molecules. The size of the gas molecule further impacts how the fragment ions are scattered and this needs to be minimised in order to assist in ion detection. Charge transfer to the gas should also usually be avoided. For these reasons, helium, argon and xenon are common collision target gases. They are all unreactive, monoatomic gases with high ionisation potentials.

4.2.2 Collisional Activation Theory

In the laboratory frame of reference, the description of a collision between an ion and a molecule involves the motion of each in three co-ordinates (x,y,z) and the internal energy of an ion can be defined as E_{lab}. To simplify the description of an ion-neutral collision, a centre of mass (CM) reference is adopted and the energy of an ion (E_{CM}) that is available for fragmentation depends on its mass (m_p) and that of the target gas (m_g) according to equation 4.2.

$$E_{CM} = E_{lab} \times m_g/(m_p + m_g) \qquad (4.2)$$

If sufficient excess internal energy is deposited into an ion during collision to break chemical bonds, the ion will fragment. Increasing the ion's initial kinetic energy and/or the mass of the target gas can increase the energy available for fragmentation. Multiple collisions with target gas molecules can also increase the internal energy of an ion to promote fragmentation, but at the same time also results in an increase in the probability of undesirable rearrangement reactions.

A principle shortcoming of the CAD process is a limitation in the amount of energy that can be deposited into a molecule. It has been calculated that as the mass of an ion increases, the internal energy it gains increases to a mass limit of approximately 1,500. Above this value the energy deposited during a collision begins to fall, leading to a practical limit for dissociation of a molecule of the order of 2,500 Da.

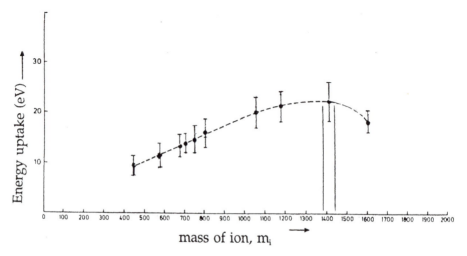

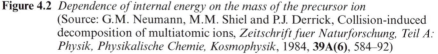

Figure 4.2 *Dependence of internal energy on the mass of the precursor ion*
(Source: G.M. Neumann, M.M. Shiel and P.J. Derrick, Collision-induced decomposition of multiatomic ions, *Zeitschrift fuer Naturforschung, Teil A: Physik, Physikalische Chemie, Kosmophysik*, 1984, **39A(6)**, 584–92)

4.2.3 High (keV) and Low Energy (eV) Collisions

The term *high energy collisions* refer to those collisions between a neutral gas molecule and a precursor ion accelerated to kinetic energies of several kiloelectronvolts (keV). This is the case in magnetic sector and time-of-flight mass spectrometers where ions leave the source with energies of typically 3–30 keV. High-energy collisions lead to the excitation of electronic states in most molecules such that their ions have a broad range of internal energies. As a consequence, most structurally viable fragmentation processes are possible.

Because ions move rapidly through the collision chamber in high-energy experiments, ions are subjected to single or only a few collisions with the target gas molecules. As a result, the mass of the target gas has a relatively small effect in high-energy CAD experiments because the centre of mass energy is a small fraction of the larger kinetic energy. This means that changes in the collision conditions (nature of the collision gas and its pressure in the collision chamber) do not result in significant changes in the MS/MS spectrum.

Low energy collisions, in contrast, occur when precursor ions have kinetic energies from a few eV up to a few hundred eV. Low-energy collisions are thought to excite vibrational states in a molecule and result in a much narrower range of internal energy distributions over high-energy collisions. A tandem mass spectrum resulting from 10 eV collisions can be dramatically different from one that results from 100 eV collisions and the nature of the target mass has a strong influence on the appearance of fragments in a low-energy CAD MS/MS spectrum. Heavier gases such as xenon and argon are often used in low-energy collision experiments to increase the probability of observing fragments. Ions undergo multiple collisions (tens to hundreds) as they pass through the chamber and internal energies are deposited in a stepwise manner and accumulated during each collision.

4.2.4 Charge Reversal and Stripping

Where an ion undergoes little to no dissociation during the collision process, it may lose electrons by *charge stripping* (equation 4.3) and undergo what is described as *charge inversion* or *charge reversal* (CR) (equation 4.4). The process of charge reversal can be useful for distinguishing the structures of ions where the CAD spectrum of the precursor ion exhibits few or no fragment ions, but its charge-reversed form readily dissociates.

$$M^{+\bullet} + N \rightarrow M^{2+\bullet} + N + e^- \qquad (4.3)$$

$$M^{-\bullet} + N \rightarrow M^{+\bullet} + N + 2e^- \qquad (4.4)$$

Collision energies of approximately 1 keV are required to effect the charge stripping of collisionally-activated ions.

Depending on the nature of the neutral gas, a charge exchange process (equation 4.5) can also occur. These reactions are exploited in *neutralisation/reionisation* (NR) experiments performed on tandem mass spectrometers. The neutralisation of positive ions is achieved using gaseous neutrals that have high ionisation efficiencies.

$$M^{+\bullet} + N \rightarrow M + N^{+\bullet} \qquad (4.5)$$

4.2.5 Photon-Induced Dissociation (PID)

Ions that contain a chromophore, or light-absorbing unit, can absorb energy in the form of photons when irradiated. Single or multiple interactions with photons can lead to the uptake of energy greater than that required to break a chemical bond. Nearly all applications of PID make use of laser light delivered to the dissociation regions through ports in the instrument by means of conventional optics of the mass spectrometer or fibre optic cables.

One of the challenges of this method is to effect the interaction of significant numbers of ions with photons within the dissociation region. Quadrupole ion trap and ion cyclotron resonance mass spectrometers are useful in this regard, since ions are trapped or held for extended periods such that repeated interactions can take place. PID, however, has been applied on both magnetic sector and quadrupole instruments usually by slowing the ions as they pass through the dissociation chamber or by passing the light along the axis of the ion beam. Unlike molecules in solution, the ions cannot lose energy to the solvent and so may remain activated for sufficient time for dissociation to occur.

PID can also be selectively administered by irradiating ions at wavelengths at which only certain components absorb the light. It can also be used to direct the fragmentation of ions, in some cases simplifying the interpretation of MS/MS spectra through the generation of fragments of a common ion series.

4.2.6 Surface-Induced Dissociation (SID)

In the mid 1980s, Cooks and co-workers developed an alternate dissociation method known as *surface-induced dissociation* (SID). In SID, ions are collided or bounced off a solid or viscous liquid surface. Ions are projected perpendicular or at some angle off the normal axis (typically 25–30°) to a planar surface, or by passing ions through narrow channels.

Ions gain internal energy from the collision and subsequently fragment. The internal energy distributions of the fragments are usually quite narrow, and the collision energy can be controlled by the initial kinetic energy of the ions, the nature of the surface and the angle of deflection.

The approach offers the potential to dissociate much larger ions than has been possible by CAD. This, however, has largely been unrealised and a significant problem with the approach stems from the degree of scattering of the ions from the surface. This can lead to an appreciable loss in ion current at the detector. Nonetheless, the approach has an advantage over CAD in that no gas is used for dissociation, thereby avoiding problems in maintaining low pressures within the mass analyser.

4.2.7 Electron Capture Dissociation (ECD)

An approach that has been successfully applied to the dissociation of large molecules, including intact proteins, is *electron capture dissociation* (ECD). Ions that pass through a high-current electron beam can be activated by ion-electron collisions.

The mechanism for this prompt fragmentation has been suggested to involve intramolecular proton transfer and ECD has been applied to sequence intact proteins by a so-called *top-down approach* in an impressive demonstration of the power of mass spectrometry. The observation that a significant proportion of backbone bonds are cleaved suggests the ECD process is *non-ergodic* (i.e. the energy is not randomised or distributed) since the localisation of a large proportion of the deposited energy is required to effect cleavage. Ions activated by collisional activation can be simultaneously subjected to ECD to enhance the level of fragmentation and thus the degree of structural detail. In the case of the protein cytochrome c (of ~12 kDa), all but 9 of the 103 amide bonds have been cleaved in activated-ion ECD experiments (Figure 4.3).

4.3 TANDEM MAGNETIC SECTOR MASS SPECTROMETERS

The first tandem mass spectrometers were those constructed of magnetic (B) and electric (E) sectors. Reverse-geometry BE mass spectrometers, in which a collision cell is located between the magnet and electric sector, were a common early form. The magnetic field strength is applied at a fixed value at any point in time leading to the transmission of ions of only one *m/z* ratio (equation 3.22). The products of CAD are then passed in turn onto the detector by scanning the electric field potential. The resulting MS/MS spectra are referred to as *mass-analysed ion kinetic energy spectra*, or MIKES.

4.3.1 Mass-Analysed Ion Kinetic Energy Spectra (MIKES)

Since an electric field separates ions according to their velocity only, the fragment ion peaks in MIKES are substantially broader than in other tandem mass spectra. This is because ions lose velocity in the direction of the precursor ion during collision. This leads to a broadening

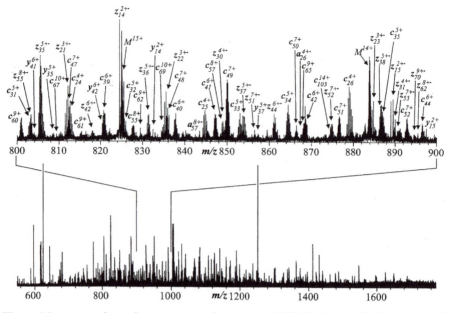

Figure 4.3 *Activated-ion electron capture dissociation MS/MS of the multiply-protonated ions of cytochrome c*
(Source: D.M. Horn, Y. Ge and F.W. McLafferty, Activated Ion Electron Capture Dissociation for Mass Spectral Sequencing of Larger (42 kDa) Proteins, *Anal. Chem.*, 2000, **72(20)**, 4778–4784)

of the ion signal over and above that associated with the distribution of ion energies from the source. A fragment ion m_f^+ formed from the unimolecular dissociation of the precursor ion m_p^+ will possess an energy of $z(m_f/m_p)eV$ where V is the accelerating potential used to extract precursor ions of charge z from the ion source. The fragment ions will be passed through the electric sector to the ion detector if the electric field strength is equal to $E_f = (m_f/m_p)E_p$. Thus MIKES are obtained by scanning the electric field from the value required to transmit the precursor ion (E_p) down to zero (or a value close to zero). The mean kinetic energy released during the formation of each of the fragment ions is measured based on the width of the peaks (at half maximum).

The peak shapes of the ion signals for the fragment ions are usually guassian in appearance. This is because each fragment ion is usually produced through a single dissociation pathway. However, where several pathways are involved and large kinetic energy releases occur, the peaks can have flat-top or dish-top shape (Figure 4.4).

To avoid broad ion signals in MS/MS spectra that lead to difficulties in assigning the correct *m/z* ratio to fragment ions, linked scans were

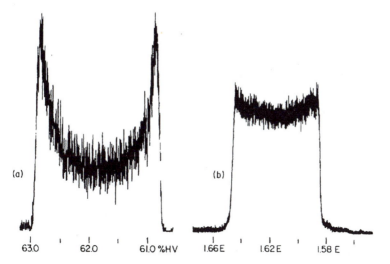

Figure 4.4 *Dishtop (a) and flattop (b) shaped ion signals evident in MIKES mass spectra* (Source: R.G. Cooks, J.H. Beynon, R.N. Caprioli and G.R. Lester, *Metastable Ions*, Elsevier, NY, 1973)

developed on two-sector mass spectrometers. These scan functions have subsequently been extended to a number of different instrument configurations.

4.3.2 Linked Scans

4.3.2.1 Fragment Ion Linked Scan. Linked scan experiments involve scanning the electric and/or magnetic fields of a mass spectrometer such that their values are always related or linked. One common linked scan is where the magnetic and electric field strengths are varied such that the ratio of their magnitudes is held at a constant value (B/E = constant). The dissociation of precursor ions is effected in a chamber immediately following the ion source on a two-sector BE or EB instrument. The initial values of B and E are those required to transmit a particular precursor ion to the detector. The B and E fields are then both successively reduced where the ratio of their magnitude (B/E) remains a constant value at all times. In this manner, all of the fragments for precursor ions of a single m/z ratio are brought into focus onto the detector. Since the fragment ions are both momentum and velocity focused the resulting MS/MS spectra achieve superior mass resolution over those obtained in MIKES. Yet since some translational energy is released during fragmentation, the mass resolution and accuracy is typically of the order of 1000. As the precursor ions are not mass-selected prior to collision, they possess a range of velocities and energies and it is typical for all isotopes of all precursor ions and fragments to be detected in the MS/MS spectrum.

The mass-selection of precursor ions, however, can be achieved in multi-sector mass spectrometers. Instruments featuring a three-sector configuration (EBE or BEE) can record MIKES type spectra in which the precursor ions are both momentum and energy focused using the first two-sectors. A collision cell preceding the last electric sector provides a region to dissociate the precursor ions. Four-sector mass spectrometers have also been constructed for high resolution MS/MS experiments in both EBEB and BEBE configurations. In these mass spectrometers, momentum and energy focusing is achieved for both the precursor and fragment ions where a dissociation chamber is positioned in the centre of the instrument between the first and last two sectors. High-resolution four-sector tandem mass spectrometers were among the first to be applied to the sequencing of entire proteins by mass spectrometry.

4.3.2.2 Precursor Ion Linked Scan.   A second common linked scan performed on magnetic-sector mass spectrometers is the so-called *precursor ion* (or B^2/E = constant) *scan*. In the case of two-sector BE or EB instruments, this scan enables the precursor ion from which a particular fragment is formed ahead of the mass analyser to be identified.

If the square of equation 3.18 is divided by equation 3.23 (see Chapter 3), one arrives at equation 4.6. Therefore, all ions that dissociate to a given fragment ion mass (*m*) are passed to the detector. This enables the precursor ion(s) from which a fragment originates to be determined.

$$B^2/E \propto m \qquad\qquad (4.6)$$

In contrast to the B/E = constant scan, both *B* and *E* field strengths are scanned proportional to the square of the ions' velocities (v^2). Hence the magnetic field does not correct for the velocity spread of the precursor ions, and information on the kinetic energy released during an ion's fragmentation is retained. As a consequence, the mass resolution in the MS/MS spectrum is degraded, but in this case it is associated with the precursor ions (not the fragment ions as in MIKES).

4.3.2.3 Neutral Loss Linked Scan.   A third linked scan useful in tandem mass spectrometry on magnetic-sector instruments is the *neutral loss linked scan*. Here, the identity of the precursor ion from which a neutral molecule was lost during dissociation is derived by using a scan in which the value for $B^2(1-E)/E^2$ is held constant. In this experiment, a fragment ion formed between the ion source and the mass analyser of a two-sector

BE or EB instrument is passed to the detector only if it differs by a constant mass from the precursor ion from which it originates. This scan is useful for identifying the nature of the analyte, since certain neutral mass losses in MS/MS spectra can be characteristic of the fragmentation of particular classes of compounds (see Appendix 5).

The linked scans described in the above sections are just some of the linked scans that can be performed on magnetic-sector mass spectrometers. A further parameter that can be scanned alone, or in a simultaneous linked fashion, is the accelerating voltage. However, since this field has a dramatic effect on the successful extraction of ions from the source (low accelerating voltages remove fewer ions), its use is more limited.

4.4 TANDEM QUADRUPOLE MASS SPECTROMETERS

Early multi-quadrupole instruments were constructed to study photodissociation processes. A triple-quadrupole (QqQ) instrument in Melbourne, Australia was the first adapted for tandem mass spectrometry in the late 1970s. The first and third quadrupoles (Q) of such an instrument are operated using a combination of RF and DC voltages. The region about the second quadrupole serves as the dissocation chamber for CAD, surface and photodissociation, and is operated in the *RF-only mode*. For this reason, it is generally denoted by the lowercase letter, q. In the case of CAD experiments, either a closed cell is situated about the second quadrupole or a more open configuration is used in which the collision gas is passed into the region of the quadrupole and maintained at an appropriate pressure (Figure 4.5).

In a tandem experiment, the RF and DC voltages of the first quadrupole are set to values that allow only ions of particular mass-to-charge ratios to be transmitted. After dissociation of these ions within the second quadrupole, the fragment ions are passed in turn to the detector by scanning the RF and DC voltages applied to the third quadrupole (see Figure 4.1).

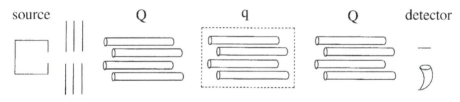

Figure 4.5 *Representation of a triple quadrupole mass spectrometer used for MS/MS experiments*

In the MS-mode, voltages applied to the first quadrupole are scanned to transmit all ions in turn through the third quadrupole to the detector.

In practice, mass discrimination effects impact the detection of some ions over others. Two modes of operation are usually employed to minimise these effects. One is a precursor ion transmission mode in which the amplitude of the RF voltage applied to the second quadrupole is set to a value that is a fraction of that required to transmit precursor ions. This results in the non-uniform transmission of fragment ions, which in the extreme may leave some products of dissociation undetected. The second fragment ion transmission mode involves scanning the RF amplitude of the second quadrupole to transmit each fragment ion with similar efficiency. It does, however, lead to a loss of precursor ions in the collision region that has a corresponding effect on fragment ion production.

Precursor ion scans can be performed on a triple quadrupole mass spectrometer by fixing the RF and DC voltages of the last quadrupole while scanning the voltages applied to the first quadrupole to transmit only fragment ions of a particular m/z ratio onto the detector. Neutral loss scans are performed by offsetting the voltages applied to the first and last quadrupole such that they transmit only fragment ions formed from precursors by a designated mass loss when scanned simultaneously.

In general, tandem mass spectrometry experiments are easier to perform on quadrupole based mass spectrometers over magnetic-sector instruments and operate at lower voltages. The former instruments, however, achieve lower mass resolution than sector instruments and are less able to exclude ions of similar m/z ratios to one another from the collision region. Triple quadrupole instruments also have found widespread use for studies of ion-molecule chemistries where a mass-selected precursor ion is studied in terms of its reactivity with a reactive gas added within the second quadrupole. Instruments featuring as many as five quadrupoles in tandem have been constructed to investigate successive reactions and also to enable MS^n (such as MS/MS/MS where $n = 3$) experiments to be performed.

4.5 TANDEM MASS SPECTROMETRY ON ION TRAPS

The MS/MS experiments described above are performed on instruments in which the mass-selection of precursor ions, their dissociation and fragment ion transmission and detection stages of tandem mass spectrometry are conducted in discrete sectors or regions of the mass spectrometer (Figure 4.1). In contrast, all stages of tandem mass spectrometry within ion traps (either quadrupole ion traps or ion cyclotron resonance instruments) are conducted within the same physical space. In these experiments, precursor ion mass-selection, dissociation and fragment ion detection events are separated over time. Ion traps offer the advantage that, since ions are more confined in a common region of the instrument during tandem mass spectrometry, a greater number of ions can be fragmented and detected.

While there are practical differences in the operation of QIT and ICR mass spectrometers for this purpose, the concepts are the same. Ions are first formed and injected into the trap or cell after acceleration and the lowering of the voltage applied to the entrance to the trap. Ions are then excited to larger trajectories by applying an RF potential. In the case of QIT's, the resonant excitation of ions is based on increasing the amplitude of the RF potential to the ring electrode, while in an ICR it depends on the RF frequency applied to opposing plates perpendicular to the ion's initial motion. These extended paths cause the ions to collide with more gas molecules already present (in the case of QIT) or added as a pulse at high pressure (in the case of the ICR) to effect collisional-activation. The fragment ions from dissociation are then ejected sequentially from the trap onto the ion detector in the case of the QIT.

4.5.1 TANDEM MASS SPECTROMETRY ON QUADRUPOLE ION TRAPS

MS/MS spectra were first reported on a quadrupole ion trap in 1987. A typical tandem mass spectrometry experiment involves:

(i)   the transmission of ions from the source into the trap where they are prevented from exiting by the voltages applied to the end caps
(ii) the selective isolation of the precursor ions by ramping the RF voltage to the ring electrode voltage above and below a particular value to store ions of a specific *m/z* ratio
(iii) the application of a resonance excitation RF voltage to the endcaps to induce faster and more extensive ion trajectories of the selected precursor ions
(iv) a lowering of the voltage applied to the end caps, with simultaneous ramping of the RF voltage applied to the ring electrode, to eject the remaining precursor ions and fragments.

The helium bath gas in the ion trap is used to both stabilise ion trajectories through collisional cooling and act as the collision gas to effect the dissociation of activated ions. Ion traps allow product ions from the first stage of an MS/MS experiment to be trapped and reactivated thereby allowing multiple stages of mass analysis and dissociation to be carried out in so-called MS^n (*e.g.* MS/MS/MS *etc.*) experiments. As many as ten stages ($n = 10$) of tandem mass spectrometry have been performed on commercial instruments, though these experiments typically have little practical application.

4.5.2 TANDEM MASS SPECTROMETRY ON FT-ICRs

Tandem experiments are performed in a similar manner on FT-ICR mass spectrometers. However, since a FT-ICR cell operates at a significantly lower pressure to that of a quadrupole ion trap, the collision gas is delivered into the cell at high pressure only during the dissociation event. The pulsing of gas in and out of the cell during tandem mass spectrometry imparts a significant burden on the instrument's pumps. To overcome this, photon-induced and electron-capture induced dissociation approaches are frequently employed since they require no such variation in the pressure within the trap.

4.6 Tandem Mass Spectrometry on TOF/TOF Instruments

Mass spectrometers consisting of two time-of-flight mass analysers have recently been developed for tandem mass spectrometry. These instruments can consist of combinations of both linear and reflecting TOF mass analysers.

Particular advantages of this type of instrument for tandem mass spectrometry are that they are relatively inexpensive to construct and also achieve high transmission and detection sensitivities. They are more limited, however, in their ability to mass select precursor ions with unit mass resolution.

In a typical tandem experiment, precursor ions have flight times in the first flight tube given by equation 3.16 (Chapter 3) when accelerated from the source with a voltage, V. Mass-selection plates are positioned ahead of the collision cell within the flight tube. A high voltage is applied to these plates to deflect all ions from the flight path except at a time when ions of a particular m/z ratio (or range) reach them. At this time, the voltage applied to the plates is switched off allowing the mass-selected ions to enter the collision cell held at voltage, V_c. Intact precursor ions leaving the cell have kinetic energies given by equation 4.7. Their time-of-flight in the second TOF analyser can be expressed by equation 3.16 where l is the length of the second flight tube.

$$1/2m_pv^2 = ze(V - V_c) + zeV_c = zeV \tag{4.7}$$

In contrast, fragment ions of mass, m_f, leaving the cell have kinetic energies given by equation 4.8.

$$1/2m_fv^2 = zem_p/m_f(V - V_c) + zeV_c \tag{4.8}$$

Their time-of-flights will be given by equation 4.9, where *l* is the length of the second TOF mass analyser.

$$t^2 = m_p/z(l^2/2eV) \ l^2/ [l + V_c/V(m_p/m_f - 1)] \tag{4.9}$$

These equations ignore contributions from the time ions spend in the deceleration and acceleration regions about the collision cell. It is usually desirable to slow ions as they pass into the cell to maximise the time available for dissociation.

A schematic representation of a tandem TOF instrument RTOF/RTOF featuring two ion reflectors is shown in Figure 4.6. These mass spectrometers are finding wider user for investigations of biological molecules, particularly in proteomics applications (see Chapter 7).

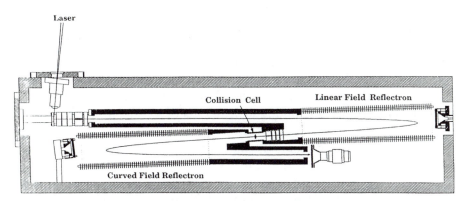

Figure 4.6 *Schematics representation of a tandem RTOF-RTOF mass spectrometer featuring both homogeneous field and "curved-field" reflectrons* (Source: T.J. Cornish and R.J. Cotter, *Rapid Commun. Mass Spectrom.*, 1994, **8**, 781–785)

4.7 TANDEM MASS SPECTROMETRY ON HYBRID INSTRUMENTS

Replacing the first TOF analyser of a TOF/TOF instrument with a double-focusing sector mass spectrometer or a quadrupole mass filter enables mass-selection to be effected with unit mass resolution. A number

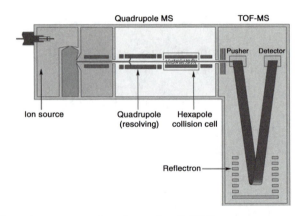

Figure 4.7 *Schematic representation of a tandem Q-TOF mass spectrometer featuring an orthogonal ESI ion source, hexapole collision cell and reflecting time-of-flight mass analyser*
(Source: Courtesy Micromass)

of hybrid instrument designs have been constructed including those of EB-TOF and Q-TOF (Figure 4.7) geometry.

Since ions are typically produced in a continuous manner from the source, it is necessary to extract packets of ions in order that time-of-flight measurements can be performed. This can result in a considerable loss in sensitivity if only small portions of the ion beam are subjected to dissociation and ultimately detected by TOF-MS. To minimise this, it is typical to project ions down the time-of-flight tube orthogonal to their initial trajectory. Since the TOF mass analyser in effect detects all fragment ions produced within a few hundred microseconds, the analysis is more sensitive than on scanning instruments that transmit only ions of a specific *m/z* value to the detector at any point in time.

In the tandem experiment, the mass-selected ions from the magnetic-sector or quadrupole analyser are decelerated prior to entering a dissociation chamber. The fragment ions are extracted using a pulsed accelerating voltage at right angles to their initial path. Since the distribution of velocities of the fragments is minimal in the direction of the TOF tube, reasonable focusing and mass resolution in the MS/MS spectrum is achieved.

Parent ion scans are performed on a Q-TOF instrument by fixing the time at which fragment ions of a particular *m/z* value reach the detector. The quadrupole is then scanned to determine values of *m/z* for the precursor ions during which fragment ions are detected.

FURTHER READING

K.L. Busch, G.L. Glish and S.A. McLuckey, *Mass Spectrometry/ Mass Spectrometry: Techniques and Applications of Tandem Mass Spectrometry*, VCH, Germany, 1988.

F.W. McLafferty (ed), *Tandem Mass Spectrometry*, Wiley, New York, 1983.

R.G. Cooks (ed), *Collision Spectroscopy*, Plenum Press, New York, 1978.

M.M. Bursey, Charge inversion of negative ions in tandem instruments, *Mass Spectrom. Rev.*, 1990, **9(5)**, 555–574.

R.A. Yost, C.G. Enke, D.C. McGilvery, D. Smith and J.D. Morrison, High efficiency collision-induced dissociation in an rf-only quadrupole, *Int. J. Mass Spectrom. Ion Phys.*, 1979, **30**, 127–136.

J.N. Louris, R.G. Cooks, J.E.P. Syka, P.E. Kelley, G.C. Stafford, J.F.J. Todd, Instrumentation, applications and energy deposition in quadrupole ion-trap tandem mass spectrometry, *Anal. Chem.*, 1987, **59(3)**, 1677–1685.

R.B. Cody, B.S. Freiser, High-resolution detection of collision-induced dissociation fragments by Fourier-transform mass spectrometry, *Anal. Chem.*, 1982, **54**, 1431–1433.

F.W. McLafferty, D.M. Horn, K. Breuker, Y. Ge, M.A. Lewis, B. Cerda, R.A. Zubarev and B.K. Carpenter, Electron capture dissociation of gaseous multiply charged ions by Fourier-transform ion cyclotron resonance, *J. Am. Soc. Mass Spectrom.*, 2001, **12**, 245–249.

Organic Mass Spectrometry

5.1 ACCURATE MASS MEASUREMENTS

The molecular weight of a compound alone can be useful for identifying an organic compound. Where this measurement is made with high mass resolution, the molecular weight of a compound can be measured to an accuracy of a few parts-per-million (ppm). The experiment is referred to as an *accurate mass measurement*.

As described in Chapter 1, Section 1.3.1, this information can be sufficient to derive the molecular formula for the compound provided that two or more possible elemental compositions are distinguishable by mass within the error of the measurement. Benzene, with an elemental formula of C_6H_6, will give rise to a singly-charged ion with a resolved monoisotopic mass of 78.0468 based on the masses for the ^{12}C and ^1H isotopes (12.0000 and 1.0078 respectively). A mass measurement of within 149 ppm would be required to distinguish this ion from that of chloropropane (C_3H_7Cl) with a monoisotopic mass of 78.0235. This value reflects one-half of the difference between the two masses divided by the mass of benzene.

5.1.1 Calibrating the Mass Scale

To measure the m/z ratio of an ion with high accuracy, it is necessary to calibrate the mass scale of the instrument by analysing a mixture of known compounds that produce ions across the mass scale. On a magnetic-based instrument, for a fixed accelerating voltage, the magnetic field strength (B) is corrected (to those values predicted from equation 3.22) to focus ions of known m/z onto the detector.

Synthetic polymers are useful calibration compounds for this purpose, since the samples are polydisperse and generate a series of ions separated in mass-to-charge by the repeating monomer unit. Compounds such as perfluorotributylamine are also commonly used since they are rich in carbon and fluorine both of which have a mass for their lightest isotope close to an integral value (arbitrarily set to 12.0000 for carbon, and 18.9984 for fluorine, see Appendix 2).

Simple metal halide salts are also used to calibrate the mass scale because they produce many *ion clusters*. These cluster ions are also separated by a repeating elemental or chemical unit. The compounds CsI and CsF are common examples. In the case of FAB ionisation, clusters observed from the matrix compound (such as glycerol) provide a convenient means with which to calibrate the mass scale using ions that appear along with those of the sample.

5.1.2 Peak Matching

Peak matching is the one of the most accurate methods used to assign the mass of an ion to a value within 1 ppm from the calculated *exact mass*. This technique is usually performed on high-resolution magnet-based instruments, but instead of adjusting the magnetic field strength, it involves rapidly switching the accelerating voltage, V. The accelerating voltage applied to ions leaving the ion source is alternately switched between two values such that the ion peaks for an unknown and reference compound are merged or overlap at the detector.

For a fixed magnetic field strength (B), it can be seen from equation 3.22 (Chapter 3) that the two ions will have the same apparent m/z ratios when the value of V is adjusted to compensate for their different mass and charge. By knowing the values of V and the m/z of ion for the reference compound with high accuracy, one can calculate the m/z ratio of the unknown compound within a small error when the accelerating voltage required to converge the ion peaks is measured. In practice, because the two ions under comparison are unlikely to have equal abundances, particularly at accelerating voltages that are not optimal, the ion signals are amplified as necessary so that their peaks can be overlapped.

Accurate mass measurements are easily achieved for relatively low molecular weight compounds of the order of a few hundred Daltons. Now that ionisation methods are available for ionising much larger molecules, further information may be required to reliably establish a compound's identity.

5.2 FRAGMENTATION OF ORGANIC MOLECULES

5.2.1 Mass Spectral Databases

Traditionally, organic molecules were analysed by EI in conjunction with a magnetic-sector mass spectrometer. Since EI is a so-called *hard ionisation* method, most mass spectra contain a series of fragment ions in addition to the molecular ion $M^{+\bullet}$ formed within the ion source. The fragmentation pattern of a compound is most influenced by the nature of the molecule, such that organic molecules generate specific signatures that have been studied over several decades.

A number of collections of EI mass spectra of organic compounds have been assembled that allow one to compare a recorded spectrum of an unknown compound with those for known compounds. One such collection is the Eight-Peak Index, published by the Royal Society of Chemistry. This collection contains in excess of 80,000 mass spectra presented as a list of the first eight major or most intense ion peaks. The spectra are classified according to the name and chemical class of the compound, the compound's molecular weight, its elemental composition, and the mass-to-charge ratios and intensities for the eight most abundant ions of the spectrum.

5.2.2 Location of Charge and Predictive Bond Fission

The initial location of a charge in a molecular ion is of particular importance in driving the fragmentation of molecules. Not all molecular ions formed from a single compound will have their charge located at the same position in the molecule, nor necessarily will that charge remain localised during the fragmentation of the compound. Yet due to the relatively high ionisation energies (typically 70 eV) used to ionise organic molecules in EI mass spectrometry, we assume from the *quasi-equilibrium theory* (Section 2.2.4) that during the formation of the molecular ion $M^{+\bullet}$ an electron can be removed from anywhere within the molecule. This ion then has sufficient time and energy prior to decomposition to transfer electrons and produce a more stable species of lower energy.

For the purpose or predicting or interpreting the dissociation of organic molecules, only those fragment ions arising from the last-formed precursors are of concern. For the most part, charge location(s) in the precursor ions drives the dissociation process. However, the products of so-called *charge remote* fragmentation processes are detected to a lesser extent, particularly in molecules such as fatty acids that contain extended aliphatic groups.

While *bond dissociation energies* (BDE) influence fragmentation, their values within ions are very different from those of neutral molecules. Thus a BDE value for a neutral molecule may not indicate that its ion dissociates along a particular path. Rather, the proximity of the charge to the dissociating bond has a greater effect on the pathway followed. As in descriptions of organic synthesis, the cleavage and formation of bonds during the dissociation of ionised organic molecules is typically described by the use of half ("fish hook") and full-headed arrows showing the direction of electron movement. As molecular ions $M^{+\bullet}$ are deficient in a single electron, bond fission (and formation) is typically represented by a homolytic process, unless evidence for a heterolytic cleavage is otherwise available.

In the following sections, fragmentations that are common across many classes of organic compounds are reviewed.

5.2.3 Homolytic Cleavage

Homolytic cleavage occurs when the electron pair of a covalent bond is transferred to two different atom centres. The site of the radical within the molecular ion $[A–B]^{+\bullet}$ is undefined in the following equations. Odd-electron ions dissociate by homolytic bond cleavage to an even-electron fragment ion and a radical (equations 5.1 and 5.2).

$$[A–B]^{+\bullet} \rightarrow A^+ + B^{\bullet} \tag{5.1}$$

$$[A–B]^{+\bullet} \rightarrow A^{\bullet} + B^+ \tag{5.2}$$

The fragment ion A^+ or B^+ with the greatest tendency to support an unpaired electron will have a higher *appearance energy*. This ion will be less stable than a low energy fragment, and it should appear with a lower relative abundance in the mass spectrum.

The decompositions of odd-electron ions are influenced by the preference for even-electron ions according to the *even-electron rule*. This states that the production of the odd-electron fragment must be accompanied by the formation of a radical species. Thus it is assumed that reactions associated with the pairing of the radical site in the molecular ion have lower energy barriers to activation.

Less favourable dissociations involving the homolytic cleavage of *two* bonds can produce odd-electron products as illustrated in equation 5.3. The bond cleaved in this example is α to the charge localised on oxygen.

$$CH_2(H)–CH_2(O^{+\bullet})H \rightarrow CH_2 = CH_2^{+\bullet} + H_2O \tag{5.3}$$

This reaction is initiated at the radical site and is driven by an electron pairing process resulting in the formation of a new bond. It is accompanied by cleavage of the α-bond. Cleavage at the β-position to the charge site is less common than α-cleavage but can be an observed fragmentation within a mass spectrometer (equation 5.4).

$$CH_3-CH_2-C(=O^{+\bullet})-CH_3 \rightarrow CH_3-CH_2^+ + CH_3-C^{\bullet} = O \qquad (5.4)$$

5.2.4 Heterolytic Cleavage

Heterolytic cleavage of a covalent bond occurs when both electrons that constitute the bond are transferred to a single atom centre. The driving force for this cleavage is the *induction effect* in which the electrons migrate to, and neutralise, the charge. Heterolytic cleavages can occur in both odd- and even-electron ions. An odd-electron ion will fragment to give a radical and an even-electron fragment ion (equation 5.5).

$$[A-B]^{+\bullet} \rightarrow A^+ + B^{\bullet} \qquad (5.5)$$

Even-electron ions dissociate to an even-electron fragment and a neutral fragment (equation 5.6).

$$[A-B]^+ \rightarrow A^+ + B \qquad (5.6)$$

5.2.5 σ-Bond Cleavage

When the bond depleted of an electron in the formation of a molecular ion $M^{+\bullet}$ is dissociated, the fragmentation is referred to as a *σ-bond cleavage* and gives rise to an even-electron product ion and a radical (equation 5.7).

$$A^{+\bullet}B \rightarrow A^+ + B^{\bullet} \qquad (5.7)$$

An example of a σ-bond cleavage is shown in equation 5.8.

$$C_6H_5^{+\bullet}C(O)-CH_3 \rightarrow C_6H_5^+ + CH_3-C^{\bullet} = O \qquad (5.8)$$

Many forms of a molecular ion may be generated during the ionisation process such that the site of the charge is not localised to a single bond. As such it can be difficult to characterise a σ-bond cleavage.

For saturated hydrocarbons, without heteroatoms that contain lone electron pairs, ionisation at a σ-bond is the lowest energy process. The bond that fragments is that which results in the production of a more substituted cation according to equation 5.9.

$$(CH_3)_3C^{+\bullet}\ CH_3 \rightarrow (CH_3)_3C^+ + CH_3^\bullet \qquad (5.9)$$

$$\overset{\times}{\rightarrow} (CH_3)_3C^\bullet + CH_3^+$$

5.2.6 Rearrangements

Fragment ions can also form by processes in which the initial bond connections in the molecular ion are reordered or arranged. Fortunately, many of these rearrangement processes have been characterised for organic molecules and therefore can be predicted based on an ion's structure. Rearrangement reactions occur with the movement of two or more sets of electron pairs.

A common rearrangement first reported by McLafferty and co-workers is referred to as the *McLafferty rearrangement*. In this homolytically-driven process, a hydrogen atom at a γ-centre migrates to the charge site through a cyclic intermediate resulting in the loss of an alkene or other stable molecule (equation 5.10).

$$CH_3-C(O^{+\bullet})-CH_2-CH_2-CH_2-H \rightarrow CH_3-C(OH^+)-CH_2-CH_2-CH_2^\bullet$$

$$\rightarrow CH_3-C(OH^+)-CH_2^\bullet + CH_2{=}CH_2 \qquad (5.10)$$

The rearrangement shown in equation 5.10 leads to the production of a *distonic ion,* namely one in which the charge and radical centres are remote from one another.

Even-electron ions can also undergo rearrangement processes during fragmentation resulting in a reorganised structure accompanied by dissociation (see for example, equation 5.11).

$$CH_3-CH(H)-CH_2-N^+H{=}CH_2 \rightarrow CH_3-CH{=}CH_2 + N^+H_2{=}CH_2 \quad (5.11)$$

The next section of this chapter reviews the major fragmentation processes that are observed in EI mass spectra for organic compounds classified by their functional groups. Compounds with more than one functional group will fragment according to processes influenced by all such groups. This section is designed to assist the reader with interpreting the EI mass spectra of organic compounds. It is not necessary to learn every single dissociation pathway for every class of compound. Rather, an understanding of the common and likely dissociation processes will aid in the interpretation of the mass spectra of any organic compound.

For a more detailed account of these processes, the reader is referred to the works of Budzikiewicz and McLafferty and their co-authors (see cited references at the end of this chapter). A summary of the fragmentation processes of organic compounds also appears in Appendix 6 at the rear of this book.

5.3 FRAGMENTATION OF ORGANIC MOLECULES BY COMPOUND CLASS

5.3.1 Hydrocarbons

The fragmentation of aliphatic hydrocarbons is usually characterised by the loss of an alkyl group and the formation of an even-electron ion (equation 5.12 and Figure 5.1).

$$[CH_3-CH_2-CH_2-CH_2-R]^{+\bullet} \rightarrow CH_3-CH_2-CH_2-CH_2^+ + R^\bullet \qquad (5.12)$$

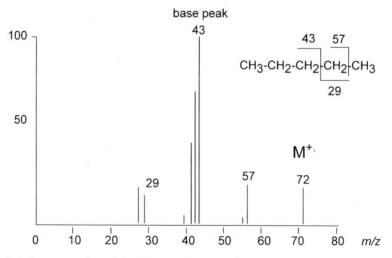

Figure 5.1 *Representation of the EI mass spectrum of pentane*

As illustrated in Section 2.2.3, fragmentation processes in which substituted carbocations are produced are favoured according to the stability of these ions ($(CH_3)_3C^+ > (CH_3)_2CH^+ > CH_3CH_2^+$).

The ability of hydrocarbons to undergo rearrangement reactions after ionisation is also well-known. Many large hydrocarbons fragment to ions corresponding to the formula $C_3H_7^+$ (*m/z* 43) (Figure 5.1) associated with an isopropyl species that subsequently loses ethylene to form the methyl cation CH_3^+ (equation 5.13).

$$CH_3\text{-}CH_2\text{-}CH_2^+ \xrightarrow{\ \ \circlearrowright\ \ } \begin{bmatrix} \overset{CH_2\text{-}CH_2}{\underset{CH_2}{\diagdown\diagup}} \end{bmatrix} H^+ \longrightarrow CH_3^+ + CH_2{=}CH_2 \qquad (5.13)$$

Hydrogen rearrangements are rarely detected in saturated hydro-carbons but are observed in the EI mass spectra of unsaturated forms. A McLafferty-type rearrangement is often encountered for unsaturated hydrocarbons with the elimination of an alkene (equation 5.14).

$$[H{-}CH_2{-}CH_2{-}CH_2{-}CH{=}CH_2]^{+\bullet} \xrightarrow{\ \ \circlearrowright\ \ }$$
$$CH_2{=}CH_2 + [CH_2{=}CH{-}CH_2]^{+\bullet} \qquad (5.14)$$

Despite these rearrangements, the fragmentations of straight-chain and branched aliphatic hydrocarbons can often be deciphered through a one-step dissociation pathway. The same is not true of cyclic hydro-carbons. Cyclohexane, for example, shows a loss of ethylene in its EI mass spectrum. Dissociation of a σ-bond cleavage followed by homolytic cleavage can account for the formation of a distonic ion at *m/z* 56 (equation 5.15) that represents the base peak in the spectrum.

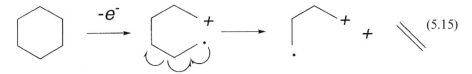

Unsaturated cyclic hydrocarbons also fragment through more complex dissociation pathways. Cyclohexene dissociates to yield major fragment ions at *m/z* 54 and 67. The latter is produced by the loss of a methyl radical after ring opening by σ-bond cleavage and hydrogen atom migration (equation 5.16). The ion at *m/z* 54 is associated with the loss of ethylene and corresponds to the ionized form of 1,4-butadiene (equation 5.17). This decomposition is a retro-Diels Alder reaction.

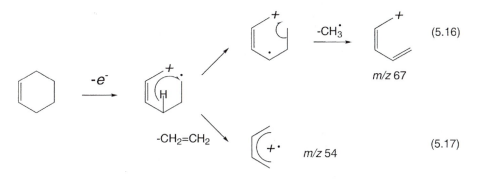

Aromatic hydrocarbons are more resistant to fragmentation and their EI mass spectra are usually dominated by the molecular ion, as well as its multiply-charged forms. In the case of benzene, in addition to the molecular ion at m/z 78 that appears as the base peak in the spectrum, other ions appear at m/z 51 and 39. The latter ion is the doubly-charged molecular ion $C_6H_6^{2+\bullet}$. Beynon and Fontaine measured the energy lost during the dissociation of $C_6H_6^{2+\bullet}$ to CH_3^+ and $C_5H_3^+$ as 2.8 eV consistent with the two charges in the parent being located over 5 Å apart. This observation suggests the doubly-charged form of benzene has an open-configuration with one of the following forms (Figure 5.2).

Figure 5.2 *Proposed structures for the ring-open configurations of the doubly-charged ion of benzene*

The fragment ion at m/z 51 can also be explained if dissociation of benzene is preceded by a ring opening reaction (equation 5.18).

$$[CH_2=CH–CH=CH–C{\equiv}CH]^{+\bullet} \rightarrow$$
$$CH_2=CH^{\bullet} + {}^+CH=CH–C{\equiv}CH \qquad (5.18)$$

Toluene dissociates by the loss of a hydrogen atom to yield a fragment ion at m/z 91. Deuterium-labelling experiments have shown that further decomposition of this ion occurs through the loss of ethylene (m/z 65) suggesting all seven hydrogen are equivalent. This supports the formation of a cyclic tropylium structure as an intermediate (equation 5.19). The proposed rearrangement mechanism for the formation of the cyclic tropylium ion is shown in equation 5.19.

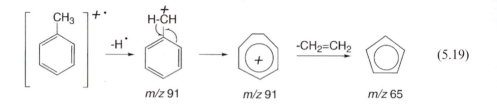

The cyclic tropylium ion often appears in the spectra of aromatic hydrocarbons, but not all alkylbenzenes necessarily give rise to a substituted tropylium ion $C_7H_6R^+$. Yet ring expansion processes have also been proposed in the fragmentation of heteroaromatic compounds. Alkylthiophenes, for example, are proposed to lose alkyl radicals through a thiapyrylium ion (equation 5.20).

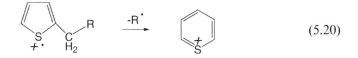

$$(5.20)$$

5.3.2 Alcohols

Aliphatic alcohols frequently give rise to EI spectra in which the ion signal for the molecular ion is generally weak or not observed. This is consistent with their lower ionisation efficiencies compared with the corresponding alkane. The absence or diminished intensity of the molecular ion is also associated with the susceptibility of alcohols to fragmention. It is convenient to visualise many of these fragmentation reactions as proceeding from a molecular ion first formed by the loss of an electron from one of the lone electron pairs from the oxygen of the hydroxyl group.

Aliphatic alcohols dissociate with the loss of a hydrogen atom, alkyl radical, water and alkenes such as ethylene. This is illustrated below where water loss proceeds through hydride ion migration to the hydroxyl group which neutralises the charge on oxygen.

$$(5.21)$$

The loss of a hydroxyl radical, in contrast to water, is rarely observed. It is common, however, to detect ions corresponding to the loss of molecular hydrogen (H_2) from the molecular ion.

Unsaturated alcohols undergo similar processes. But-3-en-1-ol, and other long chain and branched unsaturated alcohols, also dissociate following a McLafferty rearrangment according to equation 5.22.

$$(5.22)$$

The EI mass spectrum of cyclohexanol is dominated by a fragment ion at *m/z* 57 (the base peak). This ion is formed following ring opening hydrogen atom migration, and the subsequent loss of a methyl radical and ethylene.

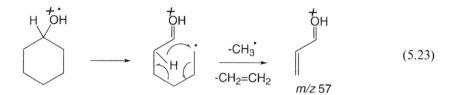

$$(5.23)$$

Many EI mass spectra of phenols have also been recorded. Like other aromatics, the spectra are dominated by intense ion signals for the molecular ion. Phenols, however, show the unique loss of carbon monoxide (–28 u) that has been proposed to occur through a cyclo-hexadienone intermediate followed by a compacting of the ring size to release CO (equation 5.24).

$$(5.24)$$

This loss is typically accompanied by the loss of 29 mass units corresponding to the CHO unit. Other substituents on the aromatic ring can stabilise the ionic products and hence can drive the fragmentation process. Alkyl substituents positioned *para* to the phenolic hydroxyl group can lose fragments from the benzylic position since the resulting ion has a stabilized oxonium ion form (Figure 5.3). An equivalent ion cannot form if the alkyl substituent is located *meta* to the phenolic hydroxyl group.

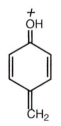

Figure 5.3 *Oxonium ion formed by the loss of alkyl substituents* para *to the phenolic hydroxyl group*

Shannon and others have studied in some detail the mass spectra of benzyl alcohols. These exhibit abundant ion signals associated with the molecular ion (m/z 108 in the case of benzyl alcohol) in addition to fragments formed by the loss of a hydrogen atom and carbon monoxide. This fragmentation has been proposed to occur through the formation of a seven-membered ring intermediate with the subsequent loss of CO to yield a benzenium ion (m/z 79) (equation 5.25). This ring can then lose molecular hydrogen to form $C_6H_5^+$ (m/z 77). An additional ion in the mass spectrum of benzyl alcohol is due to the loss of hydroxyl radical to yield the benzyl cation (m/z 91) that can be stabilised through the formation of a tropylium ion discussed earlier. A weaker fragment corresponding to the successive loss of molecular hydrogen and a hydrogen atom from the side chain gives rise to the benzoyl cation $C_6H_5CO^+$ (m/z 105) (equation 5.26).

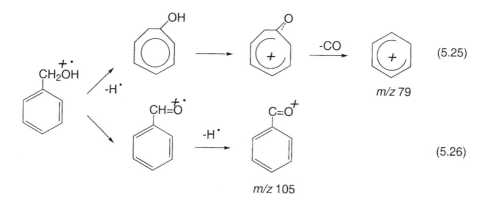

5.3.3 Ethers

The molecular ion generally appears in greater abundance in the EI mass spectra of ethers relative to their corresponding alcohols. Like alcohols, however, a common cleavage observed for ethers involves fission of a α-bond. Asymmetrical ethers can give rise to two products by this process. The more substituted ions will tend to form preferentially as illustrated in equation 5.27 over 5.28.

$$(CH_3)_2CH-O^{+\bullet}-CH_2-CH_3 \rightarrow\rightarrow (CH_3)_2CH^+ + {}^\bullet O-CH_2CH_3 \quad (5.27)$$

$$\rightarrow (CH_3)_2CH-O^\bullet + {}^+CH_2CH_3 \quad (5.28)$$

Another fragmentation process observed for ethers is hydrogen atom migration to the charge site with loss of an alkene (equation 5.29).

$$(CH_3)_2CH–O^{+\cdot}–CH_2–CH_3 \rightarrow CH_3–CH=O^+–CH_2–CH_3 + CH_3^{\cdot}$$

$$(5.29)$$

$$\rightarrow CH_3–CH=OH^+ + CH_2=CH_2$$

Unsaturated ethers, such as alkyl vinyl ethers, can undergo a McLafferty rearrangement resulting in the loss of an alkene and a stable carbonyl cation (equation 5.30).

$$[H–CH_2CH_2–O–CH=CH_2]^{+\cdot} \rightarrow$$

$$CH_2=CH_2 + [O=CH–CH_3]^{+\cdot}$$

$$(5.30)$$

Ethers possessing aromatic groups fragment through the simple α-cleavage processes shown above. Phenyl ethers also have been observed to undergo secondary fragmentations involving the loss of carbon monoxide as illustrated in equation 5.31.

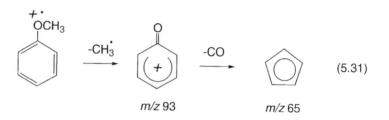

$$(5.31)$$

This fragmentation is blocked when an additional methylene group is located between the oxygen atom and the aromatic ring. Such benzyl ethers have EI mass spectra that are dominated by the benzyl cation $C_6H_5–CH_2^+$ (*m/z* 91) formed by α-cleavage.

5.3.4 Amines

Aliphatic amines generally ionise poorly by EI. However, due to the basic nature of the amino groups, stable $[M+H]^+$ ions can be produced in high yield during chemical ionization. α-Cleavage is the predominate fragmentation pathway (shown for ethyl amine in equation 5.32) with β-cleavage, and to a lesser extent γ-cleavage, becoming more important as the size of the carbon chain increases.

$$[CH_3CH_2–NH_2]^{+\cdot} \rightarrow CH_3^{\cdot} + CH_2=N^+H_2 \ (m/z\ 30)$$

$$(5.32)$$

One of the most dominant rearrangment processes observed in the spectra of aliphatic amines, arises from the transfer of an alkyl group from the carbon α to the amine group to the equivalent carbon on the opposite side followed by cleavage of the N-C bond (equation 5.33). This fragmentation process is only possible for secondary and tertiary amines.

$$[(CH_3)_2CH\text{–}NH\text{–}CH_3]^{+\bullet} \rightarrow CH_3CH\text{=}NH^+ + \text{ }^\bullet CH_2CH_3 \qquad (5.33)$$

Aromatic amines, in contrast to aliphatic amines, give rise to dominant molecular ions in their EI spectra. In many cases, as in the spectrum of aniline, the molecular ion is the base peak of the spectrum. The formation and stability of the molecular ion is attributed to electron transfer with the π-electrons of the ring (Figure 5.4).

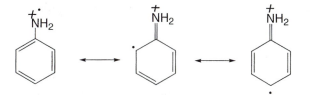

Figure 5.4 *Proposed structures for the molecular ions of aniline that contribute to its stability by electron transfer with the aromatic ring*

5.3.5 Aldehydes and Ketones

The addition of a carbonyl group to an alkane considerably lowers the ionisation energy of the molecule. The lowest energy form is that in which an electron is removed from one of the lone pairs of the carbonyl oxygen. Removal of an π-electron from the C=O bond requires more energy (typically 10.6 eV) than that required to remove an electron from a α-bond (11.5 eV). Since most EI spectra are recorded at 70 eV, all forms of the molecular ion are possible.

A common dissociation pathway in both aldehydes and ketones is that which results from cleavage of the bond α to the carbonyl group. This results in the loss of a hydrogen atom in the case of aldehydes and alkyl groups in both systems with the formation of the resonance stabilized ion R–C$^+$=O (equation 5.34).

$$[R\text{–}CO\text{–}R']^{+\bullet} \rightarrow R\text{–}C^+\text{=}O + H^\bullet \text{ or } R'' \qquad (5.34)$$

The ion H–C$^+$=O (*m/z* 29) is a signature fragment in the case of aldehydes and often appears as the base peak of the spectrum.

A McLafferty rearrangement can arise in the case of long chain aldehydes, as illustrated for the butyraldehyde ion in equation 5.35. This results in the cleavage of the β-bond through a cyclic transition state and the loss of an alkene. The mass of the latter provides an indication of the degree of branching in the molecule.

$$(5.35)$$

The aromatic ketone benzophenone is dominated by a strong molecular ion (m/z 182). The base peak of the spectrum at m/z 105 arises from the benzoyl ion C_6H_5–$C{=}O^+$ with a corresponding benzyl ion detected at m/z 77. The latter is generated as a secondary fragment of the benzoyl cation in the case of alkylphenones such as acetophenone.

5.3.6 Carboxylic Acids, Esters and Amides

The spectra of aliphatic carboxylic acids and esters, like aldehydes and ketones, are dominated by ions associated with a McLafferty rearrangement and the loss of an alkene.

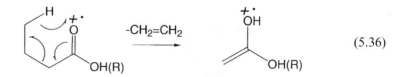

$$(5.36)$$

Both the resonance-stabilised $O{=}C^+{-}OH \leftrightarrow {}^+O{\equiv}C{-}OH$ (m/z 45) and $O{=}C^+{-}OR$ ions predominate in the EI spectra of acids and esters respectively. Cleavage β to the carbonyl group gives rise to the resonance-stabilised ions ${}^+CH_2{-}C({=}O){-}OH$ and ${}^+CH_2{-}C({=}O){-}OR$.

The loss of water from a carboxylic acid usually requires an aliphatic chain of at least four carbon atoms long with hydrogen atom transfer from the γ-position accompanied by the cleavage of C–H and C–O bonds.

The loss of a hydroxyl or alkoxide radical from a carboxylic acid or ester is favoured if the resulting product ion is stable. As an example, the EI mass spectrum of benzoic acid is dominated by fragments associated with the loss of the hydroxy radical and the subsequent loss of CO (equation 5.37).

$$[C_6H_5{-}C({=}O){-}OH]^{+\bullet} \rightarrow C_6H_5{-}C{\equiv}O^+ \rightarrow C_6H_5^+ \qquad (5.37)$$

Deuterium-labelling studies, however, have shown that the loss of HO˙ to form the ion at *m/z* 105 does not solely involve the hydrogen atom of the carboxylic acid group. The EI spectrum of *ortho*-d-benzoic acid, for example, exhibits ions at both *m/z* 105 and 106 due to what is known as the *ortho-effect*. The former product arises from transfer of a proximal hydrogen atom of the ring to the carboxylic acid group prior to hydroxyl radical loss (equation 5.38).

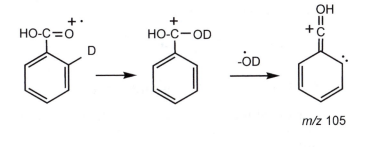

$$(5.38)$$

As one would expect, the EI mass spectra of amides resemble that of their corresponding acid and ester. An additional process, in the case of secondary and tertiary amides, results from the cleavage of the N–C bond and the transfer of one or two hydrogen atoms to produce a neutral loss (R–C(=O)–NH$_2$) and the ion R–C(=O)–NH$_3^+$.

5.3.7 Halides

The halogen atoms F, Cl, Br and I have a relatively small effect on the fragmentation processes of organic compounds. Molecular ions are detected in the EI mass spectra of halides though the proportion of charge residing at the halogen atom will vary (I>Br>Cl>F) counter to the electronegativity of the atoms. As a result, α-bond cleavage (equation 5.39) occurs preferentially adjacent to I.

$$R–CH_2–X^{+\bullet} \rightarrow R^\bullet + CH_2=X^+ \qquad (5.39)$$

The intensity of the X$^+$ ion also is observed to increase as the electronegativity of the atom decreases. Hence the I$^+$ ion appears in greater abundance than the F$^+$ ion. This is consistent with the formation of R$^+$ from the fragmentation of R–X where the loss of X˙ is observed to a greater extent in iodides and bromides, over chlorides and fluorides. This process is evident in the mass spectra of alkylhalides but not aromatic halides due to the stability of the C-X bond when the halogen is attached directly to the ring.

The abundant isotopes of some halogen atoms provide a unique signature that aids in the determination of an unknown compound. All fragments containing one chlorine atom for instance should appear as two ions differing by two mass units in an approximate ratio of 3:1 due to the natural abundance of ^{35}Cl and ^{37}Cl. The corresponding ions containing a bromine atom will appear two mass units apart in a ratio of 1:1 due to the natural abundance of ^{79}Br and ^{81}Br (see Appendix 2).

The loss of hydrogen halide is a common fragmentation pathway in the case of chlorides and fluorides (equation 5.40).

$$R-CH_2-CH_2-X^{+\bullet} \rightarrow [R-CH=CH_2]^{+\bullet} + HX \qquad (5.40)$$

5.4 QUANTITATIVE ANALYSIS OF ORGANIC COMPOUNDS

Having covered many of the common fragmentation processes observed for organic molecules with various functional groups, we turn our attention to the quantitative analysis of such compounds. Since organic compounds are used widely in prescription drugs and for agricultural, food and industrial purposes, the quantitation of such compounds in living systems, extracts and the environment is of particular importance. While the identification of an organic compound's structure provides valuable information, the absolute or relative levels of that compound in the sample may also be critical. For example, the identification of a performance-enhancing compound in the blood or urine of an athlete prior to competition may alone be sufficient to ban the athlete from competing. However, where the compound (such as a hormone) occurs naturally in the body it may be necessary to establish that a higher than usual dose has been administered. Mass spectrometry plays a major role in the quantitation of organic compounds in this, and many other, applications.

5.4.1 Role and Choice of Quantitation Standards

Quantitative analysis of any compound by mass spectrometry first requires that the detector response be calibrated as a function of the concentration of a compound at a particular set of operating conditions (*e.g.* ionisation conditions, ion source settings, tuning parameters, *etc.*). This allows the ion current detected by the mass spectrometer to be reliably correlated with the amount of compound in a particular sample. Note that this may involve a reasonably large number of measurements, since the detector response does not necessarily vary in a linear manner as the sample concentration changes across several orders of magnitude.

To perform the calibration optimally, the quantitation standards should have the same structural characteristics and be of a similar (though not the same) size to the compounds of interest. This ensures that the ionisation and detection efficiencies of the quantitation standards and the compounds under investigation are essentially identical. Ideally, the standards should be added to the sample; that is, they should be *internal quantitation standards*. This prevents any fluctuation in the performance of the instrument during the analysis of the standards, and subsequently the sample of interest, from adversely influencing the quantitation measurement. It is important that the internal standards be added to the sample at the earliest possible stage, so that they are subjected to the same potential losses prior to and during the analysis.

5.4.2 Calibration of the Detector Response

This stage of the analysis involves measuring the ion current of a particular standard or set of standard compounds as a function of their concentration. Depending on the concentration variations predicted for the samples that are to be analysed, the concentration of the standards might be varied over one or several orders of magnitude. Where the measurements are performed on a mass spectrometer featuring a scanning mass analyser (magnetic or quadrupole-based, including ion traps), the instrument is operated in the *selected ion monitoring* (SIM) mode. In this mode, the mass analyser is scanned over very small *m/z* ranges (about the ion signals of the standard(s)) to detect the majority of (if not all) the ions present. If a larger *m/z* range were scanned, the ions produced from the standard compound(s) would be passed to the detector during only part of the scan; the time the ions are detected being dependent on the scan rate of the instrument. In other words, during most of the scan the mass analyser would attempt to transmit ions of different *m/z* ratios than those of the standard(s) to the detector. In the SIM mode, many scans and thus mass spectra are obtained within a particular time interval. Where ions from several quantitation standards are to be detected, the mass analyser is scanned segmentally over several small *m/z* ranges about each ion.

Once the detector responses are measured, the area under the ion signals can be plotted as a function of the concentration of the standard(s). Any deviation in the areas obtained in subsequent runs from a line-of-best-fit establishes the error of the analysis. It is desirable to use the area under, rather than the height of, an ion signal for quantitation because the latter is highly influenced by the mass resolution of the measurement. The height of an ion peak corresponding to a particular set of isotopes is less than that where the isotopes remain unresolved.

It is typical in selecting the quantitation standards to choose compounds that have a range of molecular weights so as to produce ions across the *m/z* range of the instrument. This is because ion detectors do not detect all ions across the full *m/z* range of a particular mass analyser with equal efficiency. Low *m/z* ions are generally detected with much greater efficiency than high *m/z* ions, irrespective of whether they are produced in equal quantities in the ion source.

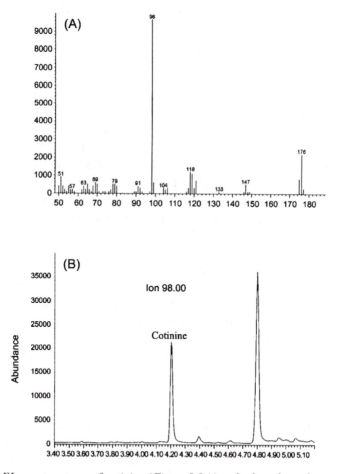

Figure 5.5 *EI mass spectrum of cotinine (Figure 5.5A) and selected ion chromatogram of fragment ion m/z 98 (Figure 5.5B) in the saliva of an active smoker*
(Source: J.-G. Kim, U.-S. Shin and H.-S. Shin, Rapid Monitoring Method of Active and Passive Smoker with Saliva Cotinine by Gas Chromatography-Mass Spectrometry, *Bull. Korean Chem. Soc.*, 2002, **23(10)**, p. 1497)

5.4.3 Quantitative Analysis of Cotinine; Example of Selected Ion Monitoring

Cotinine is a metabolite of nicotine that has been detected in both smokers and non-smokers by selected ion monitoring to assess the risk of passive exposure to cigarette smoke. The levels of cotinine were monitored in the blood, urine and saliva of both smokers and non-smokers. The EI mass spectrum of cotinine exhibits a fragment ion *m/z* 98 (the base peak) and a molecular ion at *m/z* 176 (Figure 5.5A). Selected ion monitoring of the base peak at *m/z* 98 thus affords optimal sensitivities for these experiments. A typical SIM ion chromatogram for the ion at *m/z* 98 is shown in Figure 5.5B at a cotinine concentration of 128 ng ml^{-1}.

Cotinine levels were measured based upon the ratio of the fragment ion peak area of cotinine at *m/z* 98 relative to that of the internal standard d$_3$-deuterocotinine (*m/z* 101) by interpolation from the regression line of the standard curve. Detection limits of 5–50 ng ml^{-1} were achieved among the biological matrices. The precision of the quantitation measurements was reported to be between 83.9–99.8%.

FURTHER READING

H. Budzikiewicz, C. Djerassi and D.W. Williams, *Mass Spectrometry of Organic Compounds*, John Wiley & Sons, New York, 1967.

Q.N. Porter, *Mass Spectrometry of Heterocyclic Compounds*, John Wiley & Sons, New York, 1985.

F.W. McLafferty and F. Turecek, *Interpretation of Mass Spectra*, University Science Books, 1993.

CHAPTER 6

Ion Chemistry

6.1 ELECTRON AND PROTON AFFINITIES AND MEASUREMENTS OF GAS PHASE ACIDITY

6.1.1 Electron Affinity

The most fundamental property of a molecule or atom from the perspective of mass spectrometry experiments is their ability to lose or gain electrons and form ions. The energy required for a molecule or atom to gain an electron is known as its *electron affinity* (EA). This results in the formation of a radical-anion according to equation 2.2. The electron affinity of a molecule can be defined by equation 6.1. ΔH_f^0 is the heat of formation of an molecule or ion defined as the heat absorbed or released when one mole of the entity is formed at standard temperatures and pressures (298 K = 25 °C, 1 atmosphere = 1.013×10^5 Pa). Values of ΔH_f^0 can be estimated by simple arithmetic using available thermodynamic data or by *ab initio* molecular orbital calculations.

$$\text{EA(M)} = \Delta H_f^0(\text{M}) - \Delta H_f^0(\text{M}^{-\bullet}) \tag{6.1}$$

Two types of experiments are employed to determine electron affinities. The first of these uses photon detachment methods. Using a crossed photon-molecular beam apparatus, a photon source such as a laser intersects at right angles with a beam of negative ions. The production of neutral molecules according to equation 6.2 is then studied.

$$\text{M}^{-} + h\nu \rightarrow \text{M} + \text{e}^{-} \tag{6.2}$$

The neutral molecules produced are detected using a particle multiplier. Ion cyclotron resonance (ICR) mass spectrometers are also used for these measurements where the photon beam runs parallel to the ion motion. Due to the long trapping times, the spatial overlap of the ion and photon beams is relatively large and photon detachment efficiencies are high. The minimum photon energy required to remove an electron is measured in these experiments. Photon-ion interactions are particularly suited to measurements of electron affinities since the electron binding energies of most negative ions fall in the range of 0.5 to 3.0 eV, corresponding to approximately 400 to 2500 nm.

The second type of experiment used involves a measurement of the energy required to transfer charge between a negative ion and a neutral molecule. In equation 6.3, this *charge transfer* is associated with the transfer of an electron from $A^{-\bullet}$ to B.

$$A^{-\bullet} + B \rightarrow A + B^{-\bullet} \qquad (6.3)$$

Experiments to determine whether a series of ions will transfer an electron to a particular molecule enable the electron affinity of the molecule to be bracketed.

As expected, molecules that contain electronegative atoms are more likely to bind electrons and thus will have higher electron affinities. The instability that results from pairing electrons in atomic orbitals, however, has an impact on this trend. The group XV atoms of the periodic table (nitrogen, phosphorous and arsenic) have much lower atomic electron affinities as a consequence of their half-filled *p*-orbitals.

Stabilisation of the negative charge through conjugation can raise the electron affinity of a molecule. The phenoxide radical, for example, has a higher electron affinity (2.4 eV) than other alkoxide radicals due to the delocalisation of the charge on oxygen throughout the aromatic ring.

6.1.2 Gas Phase Acidity and Proton Affinity

Measurements of electron affinities enable other thermodynamic data to be derived. For instance, the enthalpy contribution to the gas phase acidity of a compound can be derived from equation 6.4 where BDE is the *bond dissociation energy* of the M–H bond and IE is the *ionisation energy* for the molecule. The latter is defined as the energy required for the process defined by equation 2.1.

$$\Delta H_{acid}^{0}(M–H) = BDE(M–H) - EA(M) + IE(M) \qquad (6.4)$$

Equation 6.4 is derived from a thermodynamic cycle produced from a sum of the following two processes (equations 6.5 and 6.6).

$$M–H \rightarrow M + H \qquad \Delta H_f^0 = BDE(M–H) \tag{6.5}$$

$$M + H \rightarrow M^- + H^+ \qquad \Delta H_f^0 = E_i(M) - EA(M) \tag{6.6}$$

Since the primary process by which most $[M+H]^+$ ions are formed during ionisation is by the transfer of a proton from one compound to another, a molecule's gas phase acidity is an important property in mass spectrometry. Proton transfer reactions are also among the most important processes in chemical and biochemical transformations and have been studied extensively in solution.

The *gas phase acidity* of a neutral compound MH is defined as the free energy $\Delta G_{acid}^0(MH)$ required to effect the forward reaction shown in equation 6.7.

$$MH \rightleftharpoons M^- + H^+ \tag{6.7}$$

The free energy for the reaction has both enthalpy and entropy contributions according to equation 6.8.

$$\Delta G_{acid}^0(MH) = \Delta H_{acid}^0(MH) - T\Delta S_{acid}^0(MH) \tag{6.8}$$

The *proton affinity* (PA) of a neutral molecule M is defined as the energy required to effect the forward reaction shown in equation 6.9. Thus $PA(M) = - \Delta H_{acid}^0(MH^+)$.

$$M + H^+ \rightleftharpoons MH^+ \tag{6.9}$$

The traditional approach to measure the gas phase acidity of the compound MH is to react it with a base (B) and measure the degree of formation of the ion BH^+ (equation 6.10).

$$MH + B \rightleftharpoons M^- + BH^+ \tag{6.10}$$

When this measurement is performed in solution, the process is strongly influenced by the solvent medium. Proton transfer reactions with the solvent interfere with and limit the accuracy of such experiments. A mass spectrometer, by comparison, enables the intrinsic reactivity of a molecule to be studied in the absence of solvent and a number of specialised instruments for this purpose have been constructed. These instruments have led to the construction of tables of gas phase acidity data.

6.1.3 Gas Phase Acidity Measurements

Most measurements of gas phase acidities have been performed with high pressure mass spectrometers and both ion cyclotron resonance (ICR) and quadrupole ion trap (QIT) instruments. These latter instruments store ions for sufficient times to effect ion-molecule and ion-ion reactions. Reaction products and rates can be determined using these mass spectrometers.

An example of the former type of mass spectrometer is a *flowing-afterglow instrument*. These devices were originally constructed to study the chemistry of the ionosphere and feature a reaction flight tube along which reactive species can be added in a gaseous form. Their name derives from the visible glow detected within the earlier glass drift tubes from energy lost in exothermic reaction processes. The flight tube is pressurised with helium to collisionally 'cool' the ions formed in the ion source and ensure they are not in an excited (high energy) state. The ions produced in the source drift down the flight tube and react subject to the nature of the molecule added and the reaction time available. The length of the flight tube provides a reaction time domain along which rate constants can be derived.

A common method used to determine the acidity of a gaseous molecule is through bracketing methods. If a molecule MH transfers a proton to base B_1 but not to base B_2, its $\Delta G_{acid}^0(MH)$ value will be greater than $\Delta G_{acid}^0(B_1H)$ but less than $\Delta G_{acid}^0(B_2H)$. In order to rank these acidity measurements, performed on many instruments at various temperatures, on a common ΔH_{acid}^0 scale it is necessary to predict the entropy change ΔS_{acid}^0 according to equation 6.11.

$$\Delta S_{acid}^0(MH) = S^0(H^+) + S^0(M^-) - S^0(MH) \qquad (6.11)$$

For a known value of $S^0(H^+)$, the values of $S^0(M^-)$ and $S^0(MH)$ are the most dissimilar in terms of their rotational contributions. According to theory, the entropy contributions from translational, vibrational and electronic effects can be considered to be identical for M^- and MH and thus cancel each other out. The (usually small) entropy change for the reaction is then estimated from statistical mechanical considerations of rotational entropies. In practice, as these estimates can be unreliable and since most studies are conducted at a single temperature (25 °C), the small entropy contribution to a molecule's gas phase acidity is often ignored.

A problem with bracketing experiments is that the errors in known acidity data are transposed into the measured data for unknowns. An alternate method to determine gas phase acidities is the kinetic method.

6.1.4 Kinetic Method

The application of the *kinetic method* to the measurement of gas phase acidities involves measuring the ratio of product ion signals from the competitive fragmentation of the dimeric precursor ion [A . . . H$^+$ . . . B]. This ion may undergo metastable decomposition or be activated to dissociate through collision-activated dissociation (CAD) or other process. Two product ions AH$^+$ and BH$^+$ are formed with rate constants of k_A and k_B (equation 6.12).

$$[A \ldots H^+ \ldots B] \rightarrow AH^+ + B \text{ with a rate constant } k_A$$

$$\rightarrow A + BH^+ \text{ with a rate constant } k_B \qquad (6.12)$$

The relative abundances of the product ions are related to their differences in acidity $\Delta(\Delta H_{acid})$ (equation 6.13) for a proton dimer at temperature, T. R is the gas constant (8.3×10^3 J K^{-1}).

$$\ln(k_A/k_B) = \ln([AH+]/[BH+]) \approx \Delta(\Delta H_{acid})/RT \qquad (6.13)$$

A calibration curve based on the ion ratios of several species with known acidities is used to establish measured values. The kinetic method assumes that the competitive dissociations of [A . . . H$^+$ . . . B] have equal entropies and thus cancel each other out.

The popularity of the kinetic method is the result of several features including a high degree of sensitivity to differences in structure (including isotopically-labelled forms), the close agreement in values obtained by the method with those from other approaches, and the speed with which the analyses can be performed. Since the measurements are performed on a tandem mass spectrometer, it is also not necessary that the samples be pure in order for them to be studied.

6.2 ION-MOLECULE REACTIONS

6.2.1 Types of Ion-Molecule Reactions

There are many types of ion-molecule reactions including electron transfer (equation 6.3), proton transfer (equation 6.9), addition, substitution and elimination reactions. Addition reactions involve the formation of a new covalent bond, while substitution and elimination reactions are also characterised by bond cleavage. Nucleophilic substitution reactions are a class of ion-molecule reactions that have received much interest given that such processes are common in organic chemistry. In the gas-phase, in the absence of solvent, these reactions proceed along a reaction pathway with two low-energy intermediates shown in square brackets in equation 6.14.

$$N^- + M\text{–}X \rightleftharpoons [N^- \ldots M\text{–}X] \rightleftharpoons [N\text{–}M \ldots X^-] \rightarrow N\text{–}M + X^- \qquad (6.14)$$

It is convenient to present an ion-molecule reaction in terms of a *potential energy diagram* with the reaction co-ordinate on the *x*-axis. The curve (in two dimensions) or surface (in three dimensions) allows a reaction process or series of competing reaction pathways to be viewed in which the energy of the system never passes below the curve or surface.

For the reaction shown in equation 6.14, a double energy minima potential energy profile is constructed (Figure 6.1) featuring a central energy barrier that represents the transition state $[N \ldots M \ldots X]^-*$. This transition state may reside at an energy that is above or below that of the reactants.

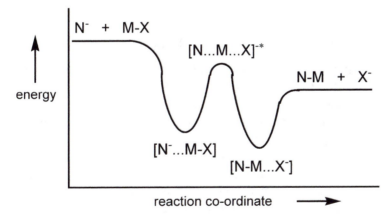

Figure 6.1 *Energy profile for an exothermic ion-molecule reaction without an activation energy barrier*

Gas-phase ion-molecule reactions importantly enable the role of solvent molecules in a chemical transformation to be explored. In the presence of solvent vapour in the mass spectrometer, the reaction of an ion with increasing levels of solvation $M^+(S)_n$ or $M^-(S)_n$ (where S is a molecule of solvent) can be explored.

Historically, most ion-molecule reactions have involved the reaction of singly-charged ions (either positively or negatively charged) and a neutral molecule though an increasing focus concerns the reaction of multiply-charged ions given their ease of production by electrospray ionisation. The reaction of multiply-charged ions with neutral molecules provides information about the site of the charges within the ion and whether such charged centres are localised or delocalised.

6.2.2 Rates of Ion-Molecule Reactions

Ion-molecule reactions are one of the fastest chemical reactions known. This is a result of attraction between the charge of the ion and the dipole of a polar molecule. This electrostatic interaction is sufficient to overcome many energy barriers to reaction products. As a result, many ion-molecule reactions are exothermic and rapidly proceed even at room temperatures. An exothermic ion-molecule reaction that proceeds without an activation energy barrier has a rate constant k that is equal to the collision-control or diffusion rate (Figure 6.1). That is, the reaction is limited only by the ability of an ion and molecule to encounter one another. As these species do so, an electrostatically-induced or natural ion-dipole interaction occurs resulting in the formation of the ion neutral complex and subsequently products. Collision-controlled rate constants are of the order of 10^{-9} cm^3 molecule^{-1} s^{-1}.

6.2.3 Ion-Neutral Intermediate Complexes

It is common when analysing most ion-molecule reactions in the gas phase to infer the formation of an intermediate complex. *Ion-neutral complexes* have been postulated for many reactions in an attempt to explain the mechanism of formation of unusual product ions observed within mass spectrometers.

Electrostatic considerations demand that when a gas-phase ion encounters a neutral species, its energy initially decreases. Thus an ion-neutral complex is formed before any energy barrier is encountered in order to generate products. In the classical S_N2 nucleophilic substitution reaction where X^- reacts with CH_3Y, the ion-neutral complex formed is represented by $[X \cdots CH_3Y]^-$ (see Figure 6.1). This results in a transition state structure denoted $[X \cdots CH_3 \cdots Y]^-$ and the formation of an ion-neutral complex for the product $[XCH_3 \cdots Y]^-$. This complex then dissociates to form the products XCH_3 and Y^-. Square brackets are used throughout.

The two components of an ion-neutral complex are associated not by a covalent bond but an ion-dipole attraction. Ion-neutral complexes can be formed from positively or negatively charged ions. The reactants are able to rotate about one another resulting in reactions that would not be geometrically possible if the partners were covalently bonded. In the gas-phase these species can be relatively long-lived with lifetimes of typically 10–100 μs. Ion-neutral complexes are particularly significant for neutral species with a small dipole moment but a large polarisability. They typically have stabilisation energies in the range of 50 kJ mol^{-1}; that is, some additional 50 kJ mol^{-1} is required for the complex to proceed to a transition state and ultimately to products.

Ion-neutral complexes lose their relevance when solvation is possible. In contrast to the reaction profile above, the S_N2 reaction between X^- and CH_3Y in solution proceeds simply from reactants to the transition structure $[X\cdots CH3\cdots Y]^-$ and then to products.

Ion-neutral complexes have been postulated as intermediates in elaborate mechanisms that attempt to explain the formation of unusual products in gas-phase ionic reactions and rearrangements. Molecular orbital calculations have also been used to implicate ion-neutral complexes in the fragmentation of certain ions. However, since these intermediates cannot be observed directly by spectroscopic methods, their existence is largely inferred rather than strictly proven. Ion-ion and ion-radical intermediate complexes have also been inferred in many gas-phase ion reactions.

6.3 KINETIC ISOTOPE EFFECTS

When the substitution of an atom in a molecule or ion by its isotope alters the reaction rate of that molecule or ion, a *kinetic isotope effect* (KIE) exists. The kinetic isotope effect is measured as the ratio of the rate constants for these reactions at a given internal energy. Where the isotopically-substituted atom is directly involved in bond dissociation or formation, the kinetic isotope effect is a primary one. When the atom is remote from the reaction centre, a secondary kinetic isotope effect is observed.

The value of the kinetic isotope effect provides important information about the particular atoms participating in a reaction, distinguishing hydrogen exchange or scrambling, identifying a stepwise reaction mechanism from a concerted one-step process, and postulating the structure for a transition state.

Most kinetic isotope effects are measured for the substitution of hydrogen with deuterium, *i.e.* k_H/k_D. The value of KIE increases from inverse ($k_H/k_D < 1$) toward normal ($k_H/k_D > 1$) as the "looseness" of the transition state increases. For reactions involving the reversible exchange of isotopes between molecular species, the kinetic isotope effect is given by the equilibrium reaction constant (equation 6.15).

$$MH^+ + A{-}D \Leftrightarrow MD^+ + A{-}H \qquad K = k_H/k_D \tag{6.15}$$

An intermolecular isotope effect exists where an isotopically-labelled ion has two possible dissociation pathways (equation 6.16).

$$D{-}CH_2CH_2Cl^{+\bullet} \to CDH{=}CH_2^{+\bullet} + HCl \tag{6.16}$$

$$\to CH_2{=}CH_2^{+\bullet} + DCl$$

Since this reaction occurs from a common precursor, the size of the isotope effect can be predicted from the relative abundances of the product ions.

FURTHER READING

M.T. Bowers (ed) *Gas Phase Ion Chemistry*, Academic Press, New York, 1979.

J.H. Futrell (ed) *Gas Phase Ion Chemistry and Mass Spectrometry*, John Wiley & Sons, New York, 1986.

R.D. Bowen, Ion-Neutral Complexes, *Acc. Chem. Res.*, 1991, **24**, 364–371.

R.G. Cooks, J.S. Patrick, T. Kotiaho and S.A. McLuckey, Thermochemical determinations by the kinetic method. *Mass Spectrom. Rev.*, 1994, **13(4)**, 287–339.

Biological Mass Spectrometry

7.1 IONISATION OF BIOMOLECULES AND BIOPOLYMERS

The low volatility and polar character of biomolecules and biopolymers initially prevented their direct ionisation and analysis by mass spectrometry. These compounds could, at best, only be studied after derivatisation of their polar groups, through methylation and acetylation, or following their degradation by, for example, acid hydrolysis. This converts large biopolymers into manageable (and ionisable) smaller molecules. Even then, only low to moderate (~1000) molecular weight compounds could be introduced into a mass spectrometer in the form of gaseous ions.

This situation changed, first with the development of plasma desorption and fast atom bombardment, and subsequently with the introduction of the electrospray and matrix-assisted laser desorption ionisation techniques. ESI and MALDI are particularly proficient at ionising large biopolymers (to several hundred thousand Daltons) without any pre-treatment or degradation of the sample. These ionisation methods are highly complementary in terms of their performance and suitability to particular samples. As a consequence, most laboratories that study biological compounds by mass spectrometry possess at least two instruments, one with an ESI source and the other with a MALDI source. Alternatively an instrument that can support both ion sources interchangeably can be used.

Peptides, proteins, glycoproteins, glycoconjugates, glycolipids, lipids, oligonucleotides and moderately-sized nucleic acids can all be efficiently introduced into a mass spectrometer by virtue of the ESI and MALDI techniques. Their importance to the analysis of biological macromolecules was recognised in 2002 by the award of the Nobel Prize in Chemistry to those who contributed to their discovery, John Fenn and Koichi Tanaka.

7.2 PEPTIDES AND PROTEINS

7.2.1 Molecular Weight Analysis

Following Barber's demonstration that peptides and small proteins could be successfully ionized by FAB in 1980, the field of protein mass spectrometry has rapidly developed. Mass spectrometry is often the first approach employed to characterise protein samples and now can provide a great deal of structural information. This includes measuring the size of a protein, its complete amino acid sequence, the nature and site of post-translational modifications and even three-dimensional structural characteristics of a protein. Recent work has led to the use of mass spectrometry for studying protein interactions.

Molecular weights are routinely measured to an accuracy of better than 0.01%, or 1 Da at a molecular weight of 10 kDa. The use of ESI and MALDI coupled to high resolution mass spectrometers, such as FT-ICR instruments (Chapter 3), has enabled the molecular weights of proteins (and other biopolymers) to be measured to an accuracy of a few ppm. This is illustrated in Figure 7.1 for the protein chondroitinase of 112 kDa measured to an accuracy of 3 Da (27 ppm). The mass accuracies obtained in all cases are far superior to those from other methods, including gel electrophoresis.

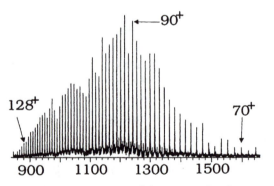

Figure 7.1 *The ESI FT-ICR mass spectrum of the protein chondroitinase of molecular weight 112 kDa measured to an accuracy of 1 Da (or 9ppm)*
*(Source: McLafferty et al., J. Am. Soc. Mass Spectrom., 1997, **8**, 380–383, Figure 1 – Part A)*

Molecular weight measurements provide useful information in their own right, and may indicate that the protein isolated from a biological sample is different or in a modified form from that expected. Heterogeneous proteins and protein mixtures are often encountered in mass spectral data even when an investigator might believe the sample to be pure. Mass spectrometry can also analyse a complex mixture of many different proteins in a single analysis and with high sample throughput, both features that are important to proteomic discoveries discussed later in this chapter.

A number of web-based algorithms have been developed to search protein databases with molecular weight information. Yet these databases contain a large number of proteins that coincidentally or due to structural similarities can share very similar, or even identical, molecular weights. As such it is often not possible to use the molecular weight of a protein alone to unequivocally identify it. For this reason, databases can be searched using a range of available information about a protein including its biological source, *pI*, molecular weight or features of its structure.

7.2.2 Mass Mapping

To assist with the identification of a protein that may already be known or display similar sequence homology to a protein that appears in a database, a *peptide mass map* can be of use. Here the protein is digested with a site-specific protease such as trypsin (which cleaves proteins on the C-terminal side of arginine and lysine residues) and the peptide products are analysed collectively by mass spectrometry (Figure 7.2). The molecular weights of these proteolytic peptides, in addition to that for the intact protein, are searched against the mass of the theoretical peptide products generated by the same enzyme for all proteins in the database. The protein(s) with the closest match based on the number and closeness of the masses matched appears in the output of the algorithm with the highest score.

A few important points must be made in relation to these searches. First, a match in the mass of a peptide with a hypothetical fragment of a known protein does not alone prove that the peptide has an identical sequence. Peptides identical in mass may coincidentally possess the same mass even when they have quite different sequences. Furthermore, some amino acid residues are indistinguishable by mass (see Appendix 8 for glutamic acid and lysine, leucine and isoleucine) and could be interchanged within a protein with no change to its molecular weight.

A second important issue concerns the level of protein coverage represented by the map. Ideally, the mass spectrum should contain ion signals for all peptides across the entire sequence of the protein. In practice, due to ionisation and detection efficiencies and the ease with which some sites are cleaved by enzymes over others, the map reflects only part of the total protein. Coverage levels vary depending on the complexity of the sample,

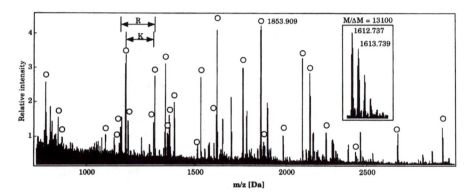

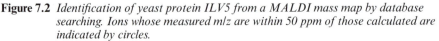

Figure 7.2 *Identification of yeast protein ILV5 from a MALDI mass map by database searching. Ions whose measured m/z are within 50 ppm of those calculated are indicated by circles.*
(Source: A. Shevchenko, O.N. Jensen, A.V. Podtelejnikov, F. Sagliocco, M. Wilm O. Vorm, P. Mortensen, A. Shevchenko, H. Boucherie and M. Mann, Linking genome and proteome by mass spectrometry: large-scale identification of yeast proteins from two dimensional gels, *Proc. Natl. Acad. Sci. USA*, 1996, **93(25)**, 14440–14445)

the nature of the proteins and the type of mass spectrometer. Thus a match of masses for peptides that only represent a portion of the total protein with those theoretically generated for a database entry leads to the possibility that they differ within other regions.

A third issue to consider when identifying proteins is mass accuracy. Irrespective of the type of mass spectrometer and measurement, some mass error is inevitable and the assignment of a protein to that of a database entry inherits these errors. This last consideration is becoming less important given that the molecular weights of peptides can be routinely measured with high accuracies. Mass measurements accurate to a few ppm are not uncommon and can be obtained even on time-of-flight instruments employing ion mirrors and time-lag focusing where appropriate mass calibration procedures are employed.

A number of web-based algorithms that use peptide mass maps to search protein databases are publicly available. These include Mascot at Matrix Science and Mass Mapper in the UK, Peptide Search at the European Molecular Biology Laboratory (EMBL), and Protein Prospector and PROWL developed in the United States. In many cases, mirror sites have been established at other laboratories to assist with data transfer throughout the world. Uniform resource locator (URL) addresses for these sites appear in Appendix 11.

An example of the output from the ProFound algorithm at the PROWL web site based on a search of the Swiss-Prot database for some of the peptide mass map data (*m/z* 1547.8, 1612.7 and 2145.1) from Figure 7.2 is shown in Figure 7.3. A mass error of 0.05 Da was selected and the cysteine residues were deemed to be unmodified. A protein molecular weight range of 20–70 kDa was chosen but no *pI* range selected.

ProFound - Search Result Summary   Version 4.10.5
The Rockefeller University Edition

Protein Candidates for search BCA7C9DF-0484-7A2DBB76 [88041 sequences searched]

Rank	Probability	Est'd Z	Protein Information and Sequence Analyse Tools (T)	%	pI	kDa	®
1	2.0e-001	0.03	T gi|6686325|sp|P71018|PLSX_BACSU FATTY ACID/PHOSPHOLIPID SYNTHESIS PROTEIN PLSX	10	5.6	35.75	®
2	2.0e-001	0.03	T gi|113783|sp|P06547|AMYB_BACCI BETA-AMYLASE PRECURSOR (1,4-ALPHA-D-GLUCAN MALTOHYDROLASE)	6	6.3	62.88	®
3	1.8e-001	0.02	T gi|401194|sp|P31015|TNA2_SYMTH TRYPTOPHANASE 2 (L-TRYPTOPHAN INDOLE-LYASE 2) (TNASE 2)	8	5.8	50.59	®
4	1.3e-001	-	T gi|120716|sp|P09316|G3P_ZYMMO GLYCERALDEHYDE 3-PHOSPHATE DEHYDROGENASE (GAPDH)	10	6.3	36.08	®
5	1.0e-001	-	T gi|124371|sp|P06168|ILV5_YEAST KETOL-ACID REDUCTOISOMERASE, MITOCHONDRIAL PRECURSOR (ACETOHYDROXY-ACID REDUCTOISOMERASE) (ALPHA-KETO-BETA-HYDROXYLACIL REDUCTOISOMERASE)	8	9.2	44.35	®
6	9.4e-002	-	T gi|1703244|sp|P53448|ALFC_CARAU FRUCTOSE-BISPHOSPHATE ALDOLASE C (BRAIN-TYPE ALDOLASE)	9	6.4	39.46	®
7	2.4e-002	-	T gi|134102|sp|P08823|RUBA_WHEAT RUBISCO SUBUNIT BINDING-PROTEIN ALPHA SUBUNIT PRECURSOR (60 KD CHAPERONIN ALPHA SUBUNIT) (CPN-60 ALPHA)	6	4.8	57.50	®
8	1.9e-003	-	T gi|1169045|sp|P43374|COX2_DEKBR CYTOCHROME C OXIDASE POLYPEPTIDE II	8	4.3	28.26	®
9	1.9e-003	-	T gi|1169041|sp|P43371|COX2_BRECU CYTOCHROME C OXIDASE POLYPEPTIDE II	8	4.3	28.28	®
10	1.8e-003	-	T gi|3123193|sp|P08897|COGS_HYPLI COLLAGENASE PRECURSOR (HYPODERMIN C) (HC)	6	4.6	28.56	®

Figure 7.3 *The ten highest-ranking entries output by the PROFOUND algorithm from a search of the SWISS-PROT database with three m/z values (m/z 1,547.8, 1,612.7 and 2,145.1) of tryptic peptide ions generated from a yeast protein ILV5*

The yeast protein IVL5 (entry P06168) appears as the fifth entry in the output. The use of a larger set of *m/z* values of ions from the mass map allows this protein to be identified with a greater confidence.

7.2.3 Peptide and Protein Sequencing

A protein can only be unequivocally identified when its entire amino acid sequence has been determined. Peptide and protein sequencing can now be accomplished solely within the confines of a mass spectrometer, or by using chemical and enzymatic approaches in conjunction with mass spectrometric analysis.

7.2.3.1 Chemical and Enzymatic Sequencing   In these approaches, the peptide or protein under investigation must be in a purified form. Other contaminating compounds in the sample can seriously compromise the analysis. The peptide or protein is treated with a chemical or enzyme to cleave amino acid residues from the N or C-terminus in a stepwise manner. An aliquot of the sample is removed at a series of time points and the collections combined. A mass spectrum is then recorded of the combined reaction products. The difference in mass of the products should correspond to the residue masses of the amino acids (Appendix 8) representing the molecular weight of an amino acid less 18 u for a molecule of water released in each step. This procedure has been described as *ladder sequencing* with the sequence of a peptide or the protein termini read directly from the mass spectrum. A partial C-terminal sequence (WCND) for an epitopic peptide of hen lysozyme has been determined based on its partial digestion with carboxypeptidase Y after reduction and alkylation of its cystein residues with ethyl pyridine (C*) (Figure 7.4).

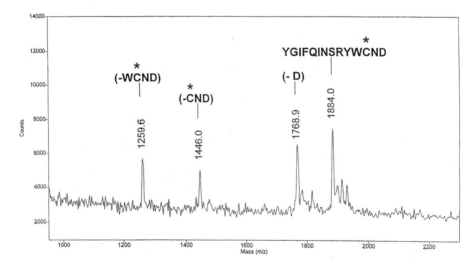

Figure 7.4 *MALDI mass spectrum of the products of limited proteolysis of an epitopic peptide of hen lysozyme with carboxypeptidase Y*
(Source: J.G. Kiselar and K.M. Downard, *Anal. Chem.*, 1999, **71**, 1792–1801, Figure 2)

Chemical methods employ the first Edman degradation reaction in which the N-terminus of the protein is converted to a phenylisothiocyanate (PITC) derivate. The modified N-terminal amino acid residue is then cleaved with trifluoroacetic acid and the process repeated. This approach is not suitable for the sequencing of N-terminally blocked proteins or peptides as the initial PITC derivative cannot be formed. The enzymatic approach alternatively makes use of amino or carboxypeptidases to cleave amino acids from the N and C-termini respectively.

In practice, the efficiency of a chemical or enzymatic cleavage decreases with the length of the polypeptide chain. Therefore, such an approach is capable of generating a complete amino acid sequence of a peptide, but not for a protein. To obtain the complete sequence for a protein, tandem mass spectrometry can be employed.

7.2.3.2 Tandem Mass Spectrometric Sequencing The sequencing of proteins by tandem mass spectrometry follows many years of study into the dissociation pathways of peptides under collisional and other ion activation conditions. Peptides have been shown to fragment along predictable pathways, for the most part involving the cleavage of the peptide backbone.

The common fragment ions for a dissociated peptide are summarised in Figure 7.5 where a nomenclature has been adopted to identify those fragments that contain the charge in the N-terminal portion (a_n, b_n, c_n, d_n) or in the C-terminal portion (x_n, y_n, z_n, v_n, w_n) of the peptide. Lower case letters are preferred so that they are not confused with nomenclature used for amino acids in their single-letter code. Many fragmentation processes involve the additional transfer of hydrogen atoms and protons but these are usually ignored for the purposes of labelling peaks in the MS/MS spectra of peptides to minimise the level of annotation. The numeral subscript (n) denotes the number of amino acid residues from either the N or C-terminus to the cleavage site. For example, a b_4 ion is formed by cleavage of the amide N–C bond at the 4th residue from the N-terminus. The fragment ions (d_n, v_n, w_n) that are formed from the cleavage of both the backbone and a sidechain group are produced only at high collision energies (keV) where precursor ions are accelerated from the ion source at kV potentials. These fragments are only observed in tandem experiments performed on magnetic sector and time-of-flight-based instruments. Although at first sight they appear to complicate a tandem MS/MS spectrum, these side-chain specific fragment ions are useful to distinguish between isobaric residues such as leucine and isoleucine that exhibit unique side chain losses.

Ideally a tandem MS/MS spectrum will contain only one series of ions in which the *m/z* value differences between each successive fragment corresponds to the mass of an amino acid residue (Appendix 8). In practice, a number of different fragment ion types can be produced simultaneously from discrete precursor ions due to the distinct structure of amino acids and the nature and energetics of the dissociation event. As certain bonds are more easily broken, and some product ions are more stable than others, the energy transferred and distributed throughout a peptide ion during activation influences its fragmentation. These factors are most pronounced in high energy (keV) dissociation experiments

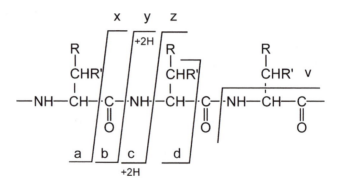

Figure 7.5 *Common fragment ions formed upon the dissociation of peptide or protein ions. Fragment ions containing the N-terminal portion of the peptide or protein are denoted a, b, c or d, those containing the C-terminal portion are denoted x, y, z, v or w. The d, v and w ions are only formed in high (keV) energy dissociation experiments*

where the nature of the fragments formed is strongly driven by the location of basic amino acid residues (particularly arginine and lysine). Low energy (eV) dissociation experiments, in contrast, generally give rise to mostly b_n and y_n type fragments regardless of the peptide sequence (see Figure 7.6).

Proteins can be sequenced following their treatment with a site-specific protease. MS/MS spectra are then acquired for each of the proteolytic peptides without the need for their purification. This establishes the sequence of segments of the protein but does not determine the order in which the segments appear in the molecule. A second protease of different specificity is used to generate a complementary set of peptides whose mass-to-charge ratios alone or in conjunction with their sequences derived from earlier MS/MS experiments allows the entire protein sequence to be assembled.

Recent advances have overcome the need to digest a protein in order to obtain sequence data. Using a FT-ICR mass spectrometer, McLafferty and colleagues have performed tandem MS/MS experiments on the multiply-charged ions of intact proteins using electron-capture dissociation. This so-called "top-down" approach has led to the production of fragment ions that cover almost all of the protein's sequence (see Figure 4.3).

The great challenge rests with interpreting the single MS/MS spectrum of the protein ions to derive the sequence. Although many of the fragment ions support more than one charge, the values for these can be measured based on the difference in the mass-to-charge ratios within their isotope distributions which are easily resolved on the FT-ICR instrument. However, since many of the fragments represent a large segment of the protein, the probability that their ions contain no ^{13}C atoms becomes exceedingly small. Thus many of the isotope distributions contain little to no detectable levels of ions containing no ^{13}C (the monoisotopic or ^{12}C-only ion peaks). It is therefore necessary to measure the mass of the fragment based on a ^{13}C-containing ion peak. The number of ^{13}C atoms in the ion must be known in order for this mass measurement to be reliable. This has been approached by comparing the ion intensities within the resolved isotopic distributions with those generated theoretically. The closest match between an experimental and theoretically calculated profile allows the ^{13}C in the ion to be assigned and the mass of the fragment to thus be derived.

These top-down experiments are an impressive demonstration of the performance of an FT-ICR mass spectrometer. The use of an electrospray ion source provides an efficient means to introduce proteins directly into the mass spectrometer. The high resolution and ion storage capabilities of the instrument coupled with electron-capture dissociation allow the protein ions to be efficiently dissociated and the fragments resolved.

7.2.3.3 Interpretation of MS/MS Spectra of Peptides   Although the methods described in this section have been developed to interpret the tandem (MS/MS) mass spectra of peptides, they can be extended to sequence proteins. Both manual and computer-assisted approaches are now in use to interpret the tandem mass spectra of peptides. The assisted approaches range from algorithms that can generate probable peptide sequences from the MS/MS data to those that attempt to identify the peptide, and the protein from which it may have been derived, by comparing the MS/MS spectral profile with a set of hypothetically-generated MS/MS spectra for all proteolytic peptides with the same mass across proteins of a database. Computer-based methods have considerably aided in the interpretation of MS/MS spectra, but it is important to note that they are fallible and programs are known to assign incorrect sequences on some occasions.

It is useful then to be practiced in interpreting MS/MS spectra of peptides by manual means. One approach to do so is illustrated for the data shown in Figure 7.6 recorded under low energy collision conditions. All mass-to-charge ratios represent monoisotopic (^{12}C only) values. The MS/MS spectrum exhibits a series of fragments from m/z 175 to 1,511 in addition to a doubly-protonated precursor ion at m/z 813.2. The monoisotopic mass of the peptide is then 1,624.4 (or 813.2 × 2 (to correct for the charge $z = 2$) − 2 (for the mass of the protons attached)). Note also

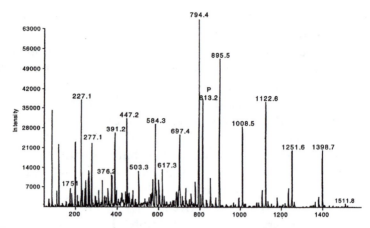

Figure 7.6 *Low energy CID tandem MS/MS spectrum of the doubly-protonated precursor (P) ions (m/z 813.2) of peptide XFENXTPXHANSR. X denotes either leucine or isoleucine*
(adapted from J.R. Chapman, ed., *Peptide and Protein Analysis by Mass Spectrometry*, Ch. 6, Fig. 6, Humana Press, NJ, 1996, p. 95)

that all of the fragment ions with $m/z > 813$ must be singly-charged since their mass is less than that of the intact peptide.

A flow chart (Figure 7.7) can be constructed where amino acid residue mass (Appendix 8) differences are searched for between one fragment ion and the next, beginning with the one with the highest m/z value. The lack of a mass match ends a branch, and another possible association must be considered. Starting at the fragment ion with the highest m/z of 1,511.8, the subtraction of 113.1 (corresponding to either leucine or isoleucine) arrives at m/z 1,398.7. Subtracting a further 147.1 units (consistent with the residue mass of phenylalanine), leads to a product ion at m/z of 1,251.6. Repeating the process further, and ignoring the precursor ion signal, the sequence of amino acids can be read in single letter code as XFENXTP where X = I or L. Two ions appear below the ion at m/z 697.4 with m/z values of 617.3 and 584.3. The first of these corresponds to a mass loss of 81 u that is inconsistent with an amino acid residue. The second corresponds to a mass loss of 113 u consistent with leucine or isoleucine. Thus we derive the partial sequence XFENXTPX.

It is now necessary to consider the fragment ions of low m/z. By repeating the subtractive process further, the sequence can be extended to XFENXTPXHANS down to the ion at m/z 175.1. This remaining mass is associated with at least one amino acid residue. If the ion at m/z 175 is a b_1 ion, the N-terminal residue must have a mass of $(175 - 1)$ (for the N-terminal hydrogen atom) or 174. Since this value is not consistent with the mass of an amino acid residue (Appendix 8), the ion

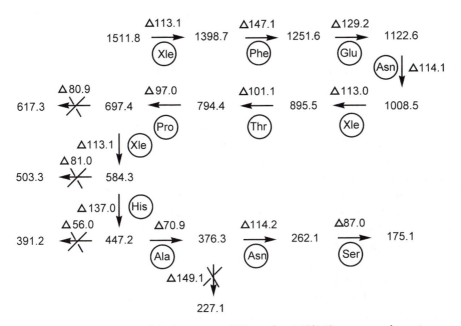

Figure 7.7 *Interpretation of the low energy CID tandem MS/MS spectrum shown in Figure 7.6*

must be considered to be a y$_1$ ion. The C-terminal residue must then have a mass of ((175 – 17) (for the mass of the C-terminal HO group) – 2) (for the mass of the two protons attached to the ion) or 156. This mass is consistent with that of arginine. For the interpretation to be correct, all ions must be of the y$_n$-series and thus the direction of the sequence from the N to C-terminus is XFENXTPXHANSR. It is not possible to distinguish leucine and isoleucine residues from MS/MS spectra recorded under low-energy conditions. These residues can only be distinguished in high-energy experiments where side chain (d$_n$, w$_n$) fragments are produced.

As illustrated above, the interpretation of tandem mass spectral data for peptides involves locating fragment ions that are separated by approximately 100 u (or strictly 57 to 186 u) consistent with the residue masses of the amino acids. Mass loss differences greater than 186 may indicate that the ions are not from the same series or type, or that one or several ions of a fragment ion series do not appear in the spectrum. For example, a spectrum may exhibit a b$_4$ and b$_6$ ion, but not a b$_5$ ion associated with cleavage of the 5th residue from the N-terminus at the amide bond. This b ion may not form due to energetic or stability issues. Under these circumstances, it is sometimes still possible to theorise as to the identity of the missing sequence based on the mass of dipeptides. It is not possible, however, without additional information to assign the order of these two amino acid residues.

7.2.3.4 Detection of Mutants and Post-Translational Modifications by MS/MS It is immediately apparent that MS/MS spectra can also be used to detect the presence of amino acid substitutions within homologous proteins (mutants) and post-translational modifications of amino acid side chains. Any such alterations will lead to a change in the *m/z* of all fragment ions that contain these amino acid residues. Fragment ions that do not contain the substituted residue or modification will appear at the same *m/z* ratio as that for the unmodified proteolytic peptide. However, post-translational modifications are often incomplete such that peptides that contain these modified residues are present at low levels in a protein digest versus their unmodified counterparts. Furthermore, some post-translational modifications (such as phosphorylation) can adversely impact the ionisation efficiency of peptides.

A number of tandem mass spectral approaches have been implemented to assist with the detection and analysis of these modified peptides and their protein counterparts. Where phosphorylated peptides are ionised and detected in the negative ion mode, a series of characteristic ions, $H_2PO_4^-$ (*m/z* 97), PO_3^- (*m/z* 79) and PO_2^- (*m/z* 63) at low *m/z* ratios are detected. Alternately, a precursor ion scan (Section 4.3.2.2) to identify all peptides that dissociate to form these fragment ions can be employed. A neutral loss scan (Section 4.3.2.3) can also be of use to detect phosphopeptides due to the characteristic loss of 98 *u* associated with H_3PO_4.

7.2.4 Protein Structure and Folding

Beyond the size and sequence of a protein, mass spectrometry can provide insights into a protein's secondary and tertiary structure. The ability to detect differences in a protein's conformational state came to light following the development of the ESI technique. Early studies of proteins showed that the charge state distribution of ions for a protein in its native state was centred at a higher *m/z* than that for the same protein in a denatured state. This was shown both as a function of pH, and more convincingly following reduction of a protein's disulphide bonds (Figure 7.8). The relationship between a protein's conformational state in the gas phase (in the absence of solvent) and that in solution is the subject of current investigation and debate. To probe the solution state of a protein, a number of indirect approaches that utilise mass spectrometry have been developed. Foremost among these is the use of hydrogen/deuterium exchange.

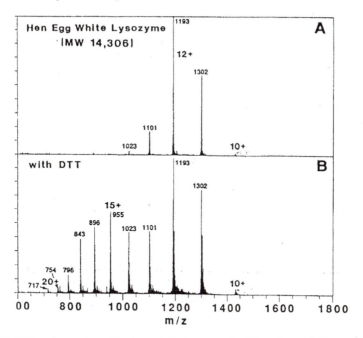

Figure 7.8 *ESI mass spectra of hen egg white lysozyme in 5% acetic acid (A) and following the addition of dithiothreitol (B)*
(Source: J.A. Loo, C.E. Edmonds, H.R. Udseth and R.D. Smith, *Anal. Chem.*, 1990, **62**, 693–698, Figure 2)

7.2.4.1 Hydrogen Exchange Mass Spectrometry   In these experiments, some of the hydrogen atoms in proteins are exchanged with deuterium or tritium by dissolving proteins in an isotopically-enriched solvent (such as deuterium oxide, D_2O). The exchange of hydrogen occurs at different rates that are dependent in part upon the accessibility of each atom to the bulk solvent. Hydrogen atoms covalently bonded to carbon undergo isotopic exchange so slowly that this exchange is not observed. Hydrogen atoms of the hydroxyl, sulphydryl, amine and carboxylic acid groups of amino acid side chains exchange very rapidly (less than a second) and their rates of exchange are indistinguishable. Hydrogen of the amide backbone, however, undergo isotopic exchange at different rates ($k \sim 5$ to 0.05 s^{-1}) that can be measured and compared. Since each amino acid, with the exception of proline, has one amide-hydrogen located in the protein backbone, hydrogen exchange levels and rates can be measured along the entire length of the protein molecule. As these backbone amide hydrogen atoms participate in formation of secondary structural elements (such as alpha helices and beta sheets), the exchange rates are affected markedly by the structure and stability of the protein. At pH 7, the half-lives for isotopic exchange of amide hydrogens within a protein may be as short as seconds or as long as several months.

The basis of the experiment is outlined in Figure 7.9. First a protein in solution at physiological pH (or some other pH of interest) is allowed to undergo hydrogen atom exchange with deuterium oxide. The reaction is then quenched by lowering the pH to 2.5 by the addition of acid and lowering the temperature to 0 °C. Under these conditions, the fast-exchanging hydrogen of the amino acid side chains are left unlabelled, while the slow-exchanging amide hydrogen atoms remain either substituted with deuterium or unexchanged. The entire protein is then analysed by mass spectrometry where the total level of deuterium incorporated is measured based on the increase in its molecular weight. For each deuterium atom that replaces hydrogen, the molecular weight of the protein increases by 1 Da. To determine the site of deuterium incorporation, a portion of the quenched sample is digested with the non-specific protease pepsin. This protease cleaves the protein efficiently at low temperature at residues across the entire protein sequence. The peptide segments are then analysed by mass spectrometry and their molecular weights measured. The level of deuterium incorporated into each peptide segment provides a way in which to measure the accessibility of the amide hydrogen to the bulk solvent across the entire protein in its original structure. Those peptides containing more deuterium represent a region of the protein backbone that is more accessible to solvent. Those peptide segments that contain minimal to no deuterium are interpreted as buried or shielded from solvent within the protein structure. In this way, the structure of a protein can be explored by mass spectrometry under a variety of solution conditions.

By extension, this approach has been used to study the interaction of proteins with other molecules. Amide hydrogen in regions of the protein within the binding site will undergo isotopic exchange at a slower rate when shielded from the bulk solvent by the interacting molecule. When such an association does not occur, the same region will exchange hydrogen with the solvent more rapidly. Thus the location of the binding site can be determined.

One difficulty with hydrogen exchange experiments arises during the analysis of the protein and peptide segments. It is critical that once the reaction is quenched or stopped that the protein does not undergo reverse exchange prior to or during its mass spectral analysis. In practice, the samples are subjected to moisture during the ionisation process that can result in deuterium loss. Therefore some care in performing the mass spectral analysis is required to ensure that the level of deuterium in

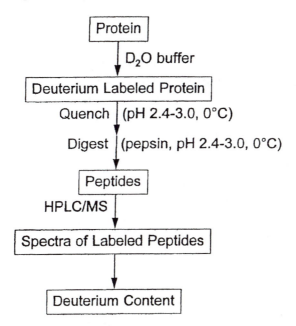

Figure 7.9 *Typical procedure employed in hydrogen exchange mass spectrometry experiments to study protein structure*
(Source: Z. Zhang and D.L. Smith, *Protein Science*, 1993, **2**, 522–531)

regions of the protein are not underestimated. One way to overcome this is to employ a reverse-exchange strategy. Here all the amide hydrogens within a protein are first completely exchanged with deuterium. This is usually achieved by denaturing the protein. The level of deuterium incorporation is then verified by mass spectrometry. The protein is then returned to its native state (usually by pH or temperature adjustment) and the reverse-exchange of deuterium with hydrogen monitored in the same manner as that described above.

Hydrogen/deuterium exchange has also been used to investigate the structure of proteins and their transient intermediate states *in vacuo* following the trapping of their ions for extended periods. These studies allow the conformational characteristics and dynamics of gaseous ions to be explored and compared with solution state observations. At least six different intermediate states for the protein, cytochrome c, have been characterised in such gas-phase experiments.

7.2.4.2 Ion Mobility Mass Spectrometry Mass spectrometry can also be employed to study the conformational characteristics of gas phase ions by measuring their mobility as they pass through an inert buffer gas. In these *ion mobility measurements*, protein ions have drift times that depend upon their average cross-section and hence conformation.

Different drift times have been recorded for the ions of proteins in their native (oxidised or disulphide-bridged) and denatured (reduced) forms.

7.2.4.3 Radical-Based Studies of Protein Structure A recent development that has been employed to study the structure and dynamics of proteins and their interactions by mass spectrometry involves their reaction with radicals. Although the reaction of radicals with proteins *in vivo* leads to their structural degradation and aggregation through cross-linking, it has been found that proteins can undergo limited oxidation without structural change when the reaction times are kept very short (several milliseconds).

Furthermore, the degree to which oxidation occurs is highly dependent on the accessibility of amino acid side chains to the bulk solvent. By measuring the site and degree of oxidation at amino acid markers throughout the protein by mass spectrometry, a protein's structure can be verified or even predicted.

The approach has a number of advantages over hydrogen exchange experiments in that the experiments can be performed extremely rapidly, the radical-induced oxidation reactions are irreversible, and the reaction timescale is sufficiently short to allow some protein conformational changes to be followed at both a global and local level (Figure 7.10). The approach has also been applied to study the dynamics of protein folding and the interactions of proteins with other molecules.

7.2.5 Protein Complexes and Assemblies

Beyond the indirect approaches described above, the development of ESI-MS for the study of proteins gave rise to early observations in which ions corresponding to intact protein complexes were sometimes detected within the mass spectrometer. Depending on the solution conditions under which the sample is introduced, and those of the mass spectrometer itself, these complexes can reflect either solution-state or non-specific associations. The pH of the solution, temperatures within the ion source, and the degree to which ions are accelerated within and as they leave the ion source, are all of importance for the detection of gas phase protein complexes.

The nature of these gas phase protein complexes is a matter of immediate question, since their molecular weights reflect that they typically have no molecules of solvent attached. Electrostatic interactions between charged groups rather than hydrophobic interactions between the component ions and the solvent are believed to play a more

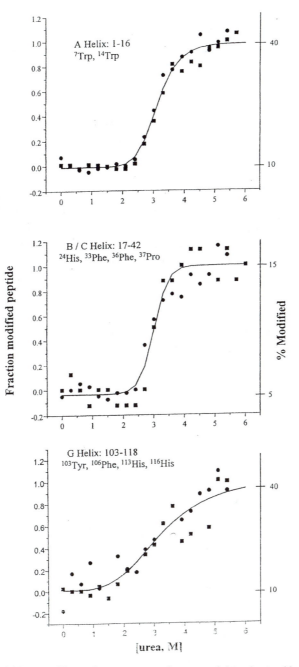

Figure 7.10 *Unfolding profiles within segments of apomyoglobin obtained by limited oxidation of the protein with hydroxyl radicals on millisecond timescales at increasing concentrations of urea*
(Source: S.D. Maleknia and K.M. Downard, *Eur. J. Biochem.*, 2001, **268**, 5578–5588, Figure 6)

important role in their stability. It cannot therefore be assumed, even when a known solution-based protein complex is observed in the gas phase, that the complex maintains the same structural features. Nonetheless, a growing body of data now suggests that relationships between solution state and gas phase protein complexes *do* exist in some systems, and that their direct detection by mass spectrometry may provide a useful means by which to identify such associations in the first instance. This has important ramifications for studies in proteomics, an ultimate description of which requires all protein associations within a cell to be identified.

Although the majority of proteins, and other macromolecular complexes, have been detected using ESI mass spectrometry (Figure 7.11), some complexes and aggregates have also been detected using the MALDI approach. This is even more surprising since samples are introduced into the mass spectrometer from a solid surface onto which the analyte is added to a high concentration of an organic matrix. Such conditions are expected to dissociate any protein or macromolecular complex prior to the ionisation event. A number of protein and other macromolecular complexes, however, have been both preserved and detected by MALDI mass spectrometry. Among them are immune complexes between protein antigens and monoclonal antibodies (Figure 7.12).

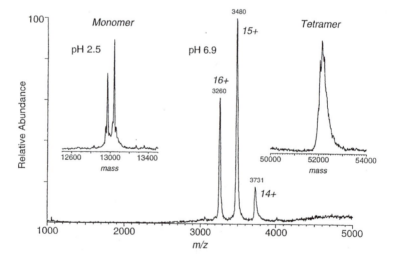

Figure 7.11 *ESI mass spectrum of streptavidin in 10 mM ammonium acetate (pH 6.9) at a concentration of 5 μm. Inserts show deconvoluted (by molecular weight) mass spectra at pH 2.5 and 6.9.*
(Source: J.A. Loo and K.A. Sannes-Lowery, in *Mass Spectrometry of Biological Materials*, 2nd edn, B.S. Larsen and C.N. McEwen (ed), Marcel Dekker, New York 1998, p. 358, Figure 5)

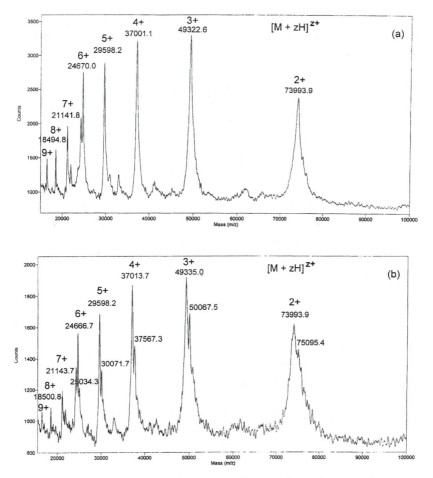

Figure 7.12 *MALDI mass spectra of the tryptic digest of the viral proteins from a type A influenza strain (a) before, and (b) after interaction with a monoclonal antibody raised to one of the proteins*
(Source: J.G. Kiselar and K.M. Downard, *J. Am. Soc. Mass Spectrom.*, 2000, **11**, 746–750, Figure 1)

An antigenic peptide that represents a surface domain of the hemagglutinin antigen of a type A influenza strain is found to preferentially bind to the antibody. Figure 7.12 shows the MALDI mass spectrum for the intact antibody before and after treatment with a mixture of peptides containing the hemagglutinin epitope. The molecular weight change that is evident in the additional ion signals associated with the antibody-peptide complex corresponds to the peptide antigen (MW 2210 Da).

A computer algorithm, known as COMPLX, has been recently developed which enables bimolecular protein and other macromolecular complexes to be identified in an automated manner from both ESI and MALDI mass spectra of the type shown in Figure 7.12. The program is of particular value for spectra that exhibit many hundreds of ion signals such that a manual interpretation of the data would be difficult or impossible. The ability to preserve protein complexes within a mass spectrometer in some circumstances is of potential value for identifying such associations in biological extracts.

Beyond such direct observations, a number of indirect approaches have made use of mass spectrometry to study protein complexes. Among these are limited proteolysis of protein complexes to release non-binding domains, hydrogen exchange studies, and the use of microfilters, chromatographic and native electrophoretic methods to separate or isolate large macromolecular complexes for their further characterisation by mass spectrometry.

7.2.6 Proteomics

The development of mass spectrometry for the study of proteins has led to its central role in *proteomics*. Proteomics, or proteome analysis, involves the identification and characterisation of the entire complement of proteins expressed in a single cell or tissue at any given point in time. *Functional proteomics* seeks to identify the protein components that are unique to diseased cells or tissue, those produced only in response to a genetic abnormality, or those of importance to a particular biological event or process.

Two primary methods are used to partially resolve the protein complement of a cell or tissue prior to mass spectrometric detection. The first of these is multi-dimensional chromatography in which proteins are partitioned according to their ionic character, molecular identity, size, and hydrophilicity using ion exchange, affinity, molecular exclusion and reverse-phase chromatography. These partitioned proteins, or their proteolytic products, are then introduced directly into an ESI-based mass spectrometer where they are characterised by molecular weight and sequence.

A second alternative strategy utilises two-dimensional gel electrophoresis to separate and display the proteins. Proteins isolated from the cell or tissue are separated in 2D gel electrophoresis in the first dimension according to their charge and in the second dimension by size. The proteins are typically digested in-gel to release their proteolytic peptides. These components are analysed by mass spectrometry in both MS and

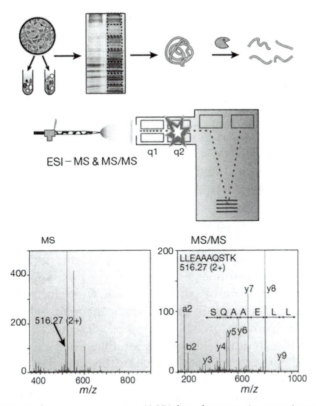

Figure 7.13 *A typical mass spectrometry (MS)-based proteomics experiment. Proteins recovered from cells are partially separated by SDS-PAGE, excised or digested in-gel, and the proteolytic peptides analysed on a tandem mass spectrometer. A Q-TOF hybrid tandem mass spectrometer is shown here (see Chapter 4, Section 4.7)*
(Source: adapted from R. Aebersold and M. Mann, Mass spectrometry-based proteomics, *Nature*, 2003, **422**, 198–207)

MS/MS experiments (Figure 7.13). Either an ESI or MALDI-based mass spectrometer can be used in these experiments.

Proteins are identified by mass spectrometry in the same manner as for single proteins. However, the use of protein and nucleotide databases are of particular importance to these investigations in order that previously characterised proteins, or their homologues, are rapidly identified. Mass map data as well as tandem MS/MS spectra of proteolytic peptides can be used to search a database for a known protein. Where tandem mass spectra are used, algorithms such as SEQUEST are capable of comparing a fragment ion profile with those for any hypothetical proteolytic peptide obtained from a known protein's sequence.

A major challenge confronted in proteomics is the management of large amounts of diverse data. Proteome analysis requires that data collected for each protein from within a cell or tissue be collated in terms of its recovery, electrophoretic or chromatographic profile, biochemical treatment, mass spectral appearance and bioinformatics discovery. Advanced computer-based bioinformatics software and systems are used for this purpose.

The field of proteomics has rapidly advanced due to the high performance and continual development of today's mass spectrometers. The discovery of new, previously uncharacterised proteins is continuing at a rapid rate. Beyond their identification at the molecular weight and sequence level, the quantitation of protein components in cellular lysates and biological extracts is also a core goal of proteomics. This represents a greater analytical challenge but one that has been addressed in part based on chromatographic detector responses, the image analysis of stained proteins on two-dimensional gels, and by mass spectrometry.

The chemical treatment of samples isolated from normal and diseased cellular extracts has been used to quantitate the relative levels of protein in each sample. In one protocol, an isotopically-enriched tag is reacted with the cysteine residue side-chains of one sample, while the same unlabelled reagent is reacted with those of the second. The two samples are then mixed and affinity chromatography is used to recover the tagged proteins. Mass spectra are then recorded for the recovered proteins where two ion signals (for the labelled and unlabelled forms) are mass resolved and detected. The relative area for these ion signals provides a quantititative measure of the levels of protein in each sample (Figure 7.14). Disadvantages of such an approach include the need to couple the reagent to proteins in each sample with equal efficiency and the ability to recover the proteins by affinity chromatography.

A particularly powerful application of MALDI-MS in proteomics is that involving a direct analysis of the spatial organisation (to a resolution of some 50 μm) of peptides and proteins in mammalian tissue sections. Caprioli and colleagues have demonstrated that it is possible to laser ablate compounds directly from a tissue section, or a blot of the tissue slice, to produce ions that span the range from m/z 1,000 to 100,000. Two-dimensional maps can then be reconstructed based on the intensity of these ions to provide, to a first approximation, the relative levels of certain molecules within the tissue (Figure 7.15). This mass spectrometric "imaging" of proteins and other components from tissue is still in its infancy but if well-developed would provide a powerful and rapid technique for use in both research and clinical settings.

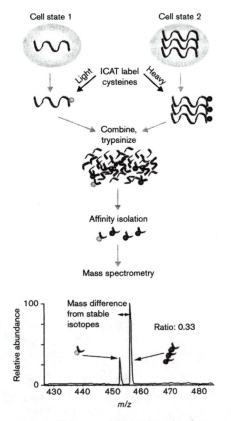

Figure 7.14 *The isotope-coded affinity tag (ICAT) approach to quantitate the relative levels of proteins from cellular extracts by mass spectrometry*
(Source: adapted from S.P. Gygi, R. Aebersold and M. Mann, Mass spectrometry and proteomics, *Curr. Opinion in Chem. Biol.*, 2000, Vol. 4, 489–494, Figure 2)

In a similar manner, MALDI mass spectrometry has been used to obtain protein profiles of unfractionated microorganisms including viruses, bacterial and fungal cells, and spores. The positive and negative MALDI mass spectra of proteins desorbed directly from the bacteria *Helicobacter pylori*, where 26995 intact cells were introduced into the mass spectrometer, are shown in Figure 7.16.

Such spectra have also been generated using laser ablation mass spectrometry where airborne microorganisms are introduced into the mass spectrometer as an aerosol. Microorganisms may also be identified, based on an analysis of the proteolytic peptides generated from their proteins in a mass map experiment.

A future goal of proteomics is to characterise the association of proteins *en masse* in cells and tissues. This will allow studies of

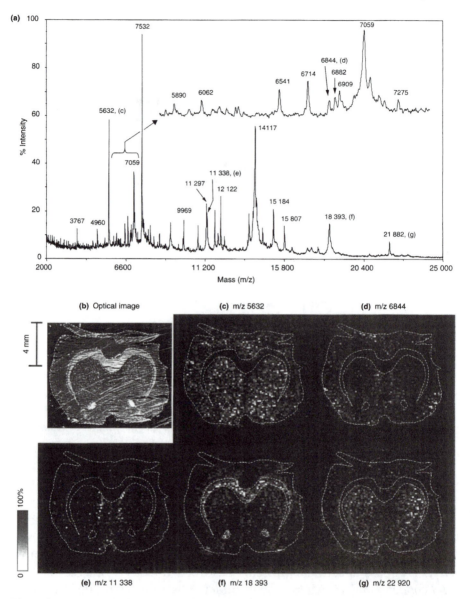

Figure 7.15 *MALDI-MS imaging of a transveral section of rat brain. (a) Survey profile taken randomly across the section, (b) optical image of the section before matrix application, (c)–(g) ion density maps obtained at different m/z values* (Source: P. Chaurand, S.A. Schwartz and R.M. Caprioli, Imaging mass spectrometry: a new tool to investigate the spatial organization of peptides and proteins in mammalian tissue sections, *Current Opinion in Chemical Biology*, 2002, **6(5)**, 676–681, Figure 3)

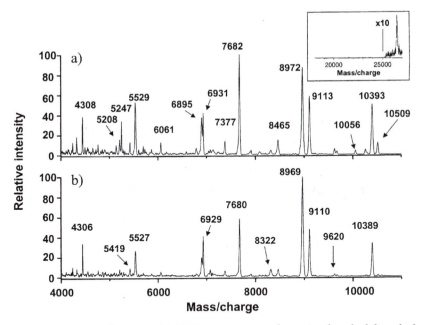

Figure 7.16 *Positive and negative MALDI mass spectra of proteins desorbed directly from the bacteria* Helicobacter pylori
(Source: P.A. Demirev, J.S. Lin, F.J. Pineda and C. Fenselau, Bioinformatics and mass spectrometry for microorganism identification: Proteome-wide post-translational modifications and database search algorithms for characterization of intact *H. pylori*, *Anal. Chem.*, 2001, **73(19)**, 4566–4573, Figure 1)

protein function to be followed on a global scale. Current genetic and bio-chemical methods are sure to be supplemented by the use of mass spectrometry for these endeavours, particularly since mass spectrometry is already in widespread use for the study of protein structure and interactions.

The use of MALDI mass spectrometry in this regard has been shown to be able to survey the structure and antigenic identity of the influenza virus. This has been achieved without the need to immobilise either the viral antigens or their cross-interacting antibodies. It has been found that the interaction between an antibody and a specific region (epitope) of a protein antigen can be preserved on a MALDI surface from which all non-binding peptides, generated after proteolysis, can be preferentially ionised. A careful comparison of the MALDI mass spectrum of the proteolysis products of an unreacted antigen mixture (the control) versus that of the antibody-containing mixture enables binding domains to be identified and characterised. Such spectra for the tryptic digest of all four viral antigens from a type A influenza strain are shown in Figure 7.17. A measure of the ion abundances in both spectra enable an

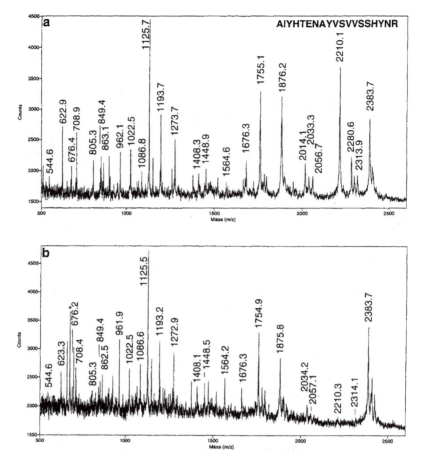

Figure 7.17 *MALDI mass spectra of the tryptic digest of the viral proteins from a type A influenza strain (a) before and (b) after reaction with monoclonal antibody. The peak labelled with an asterisk is attributed to a matrix cluster ion* (Source: J.G. Kiselar and K.M. Downard, *Biochemistry*, 1999, **38**, 14185–14191, Figure 2)

antibody-binding domain (m/z 2,210.1) to be identified within a single (hemagglutinin) antigen based upon its reduced intensity.

7.3 OLIGONUCLEOTIDES AND NUCLEIC ACIDS

Like peptides, oligonucleotides are linear polymers comprised of nucleoside monomers linked by a phosphodiester group. Conveniently, these polymers are composed of just four different natural monomer units: adenosine (A), guanosine (G), cytidine (C) and uridine (U) in the case of ribonucleic acids (RNA), and the deoxy forms of A,G,C and thymidine

(T) in the case of deoxyribonucleic acids (DNA) (see Appendix 9). This would appear to make the identification and sequencing of nucleic acids by mass spectrometry more straightforward than comparably sized proteins. In reality, however, oligonucleotides and nucleic acids are more fragile within a mass spectrometer than proteins and prone to degradation at the phosphodiester linking group. The larger mass of each of the mononucleotides also impacts the ionisation efficiency of oligonucleotides and nucleic acids and the ability to induce their dissociation in tandem mass spectrometry experiments. Nonetheless, mass spectrometry has been applied to the molecular weight and sequence analysis of oligonucleotides and nucleic acids.

7.3.1 Identification of Modified Nucleosides

Post-transcriptional and other cellular processes result in a wide range of structural modifications occurring at the purine and pyrimidine bases of RNA. Mass spectrometry has played a significant role in the identification of these modified bases, originally through the enzymatic hydrolysis of RNA. Such hydrolysates are analysed by either LC-ESI-MS, or GC-MS in the form of their volatile trimethylsilylated derivatives. The mass difference between the nucleoside and natural forms provides information concerning the type of modification(s). Tandem (MS/MS) mass spectrometry of the modified nucleosides can also be employed to enable the site of the modifications to be determined.

Although both RNA and DNA segments in excess of 100 bases have been successfully detected within mass spectrometers, they are typically ionised with less efficiency than proteins of a comparable size. This is associated with the fact that the negatively charged phosphodiester group shows both a propensity to dissociate and to adduct alkali metal cations such as Na^+ and K^+. These traits adversely impact the study of the low amounts of DNA and RNA that is usually available. As a consequence, mass spectrometry is yet to compete with some other analytical approaches for the detection and characterisation of nucleic acids. It does, however, offer a highly complementary approach and has proved useful for certain applications.

7.3.2 Sequencing of Oligonucleotides by Tandem Mass Spectrometry

High-resolution mass spectrometry can be used to verify the sequence of a synthetic or isolated RNA and DNA segment. To avoid contributions to their mass from metal ions, samples are desalted prior to analysis.

Like proteins, RNA and DNA can be digested into smaller oligo-nucleotides that are more amenable to analysis by the use of restriction endonucleases. These segments and synthetic oligonucleotides can be sequenced in the same manner as described for peptides (see Section 7.2.3.2). Tandem mass spectrometry has been applied to sequence simple short-chain oligonucleotides and a nomenclature has been proposed to describe the observed fragments (Figure 7.18).

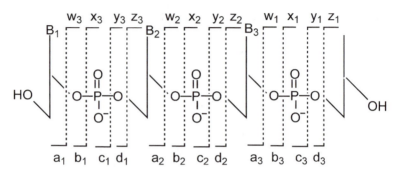

Figure 7.18 *Nomenclature to designate fragment ions detected upon the dissociation of oligonucleotides in a mass spectrometer*

The ESI tandem CID spectrum of oligonucleotide d(GCTGGCATC-CGT) recorded in the negative ion mode is shown in Figure 7.19. Although MS/MS experiments are useful for particular applications, they fail to compete with automated high-throughput sequencing by means of the polymerase chain reaction (PCR).

7.4 OLIGOSACCHARIDES AND GLYCOCONJUGATES

Oligosaccharides and glycoconjugates are composed of monosaccharides such as glucose, hexose and fructose linked through glycosidic bonds. Determining the structure of these compounds is far more difficult than is the case for peptides and oligonucleotides. This is due to the isomeric nature of some monosaccharides and the fact that these monomers can couple in a multitude of ways to produce highly branched structures.

Like oligonucleotides, oligosaccharides and glyconjugates also ionise less efficiently than their equivalently sized peptide and protein counter-parts. It is often necessary to analyse oligosaccharides in the negative ion mode due to the propensity of free hydroxyl groups to support a negative charge. Alkali metal ions often present in a sample may coordinate to these groups resulting in an unpredictable increase in the mass of the compound. To prevent this, glyconjugates can be methylated, acetylated or otherwise derivatised to facilitate their ionisation as positive ions

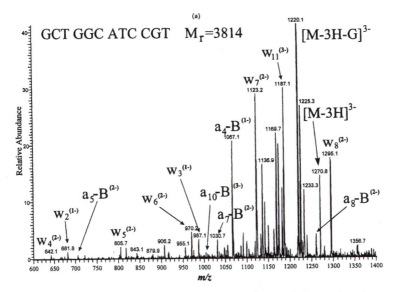

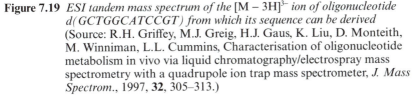

Figure 7.19 *ESI tandem mass spectrum of the* [M − 3H]³⁻ *ion of oligonucleotide d(GCTGGCATCCGT) from which its sequence can be derived* (Source: R.H. Griffey, M.J. Greig, H.J. Gaus, K. Liu, D. Monteith, M. Winniman, L.L. Cummins, Characterisation of oligonucleotide metabolism in vivo via liquid chromatography/electrospray mass spectrometry with a quadrupole ion trap mass spectrometer, *J. Mass Spectrom.*, 1997, **32**, 305–313.)

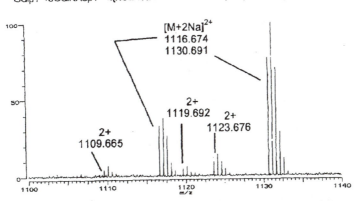

Figure 7.20 *Molecular ion region of the ESI FT-ICR mass spectrum of permethylated ganglioside* G_{D1b} (Source: C.E. Costello, Bioanalytic applications of mass spectrometry, *Current Opinion in Biotechnology*, 1999, **10(1)**, 22–28, Figure 2a)

(Figure 7.20). This method of analysis proved mandatory prior to the development of the newer desorption ionisation methods (such as FAB and MALDI) in order to volatilise oligosaccharides.

Heterogeneity is often encountered in glyconjugate samples where individual components differ in molecular weight by that of a mono-saccharide (Figure 7.21). A molecular weight profile of the sample provides an immediate indication as to its complexity and may identify potential heterogeneity among the components. Hydrolysis of the sample prior to analysis can allow the component monosaccharides to be identified though care must be taken to avoid modifying groups being removed in this process. Glycosidases are also used to evaluate such samples as described later in section 7.4.2.

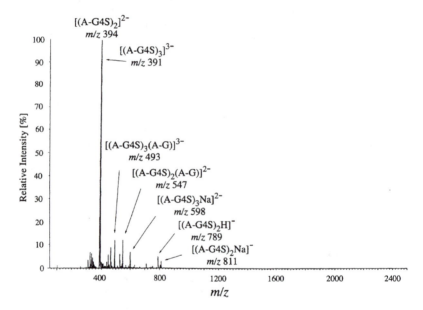

Figure 7.21 *ESI mass spectrum of the enzymic hydrolysate of kappa-carrageenan*
  containing ions associated with the tetrasaccharide (A-G4S)$_2$ and
  hexasaccharide (A-G4S)$_3$
  (Source: D. Ekeberg, S.H. Knutsen and M. Sletmoen, Negative-ion
  electrospray ionisation-mass spectrometry (ESI-MS) as a tool for analysing
  structural heterogeneity in kappa-carrageenan oligosaccharides,
  Carboyhydrate Research, 2001, **334(1)**, 49–59, Figure 4)

7.4.1 Sequencing of Oligosaccharides by Tandem Mass Spectrometry

The dissociation of oligosaccharides within a tandem mass spectrometer results primarily in the cleavage of the glycosidic bonds. A nomenclature has been proposed to designate the possible fragment ions formed by the dissociation about glycosidic bonds in which charge is retained at either the reducing or non-reducing terminus resulting in the production of B, C, Y and Z ions (Figure 7.22). A numbered subscript identifies the number of monomeric units toward the termini where the B and C ions contain the non-reducing terminus and Y and Z ions contain the reducing terminus. The Greek letters (α, β, *etc.*) are used to denote the branch position by a subscript. Two additional fragments A and X are a result of ring cleavage. These ions are designated with numerical superscripts to denote the bonds broken within the ring. For instance, the ion nomenclature $^{1,3}A_{2\alpha}$ denotes a fragment formed from the cleavage of the first and third bonds of the monosaccharide unit (in a clockwise direction from the O-C bond, denoted bond zero) at the first two sugars from the non-reducing terminus along the first (α) branch.

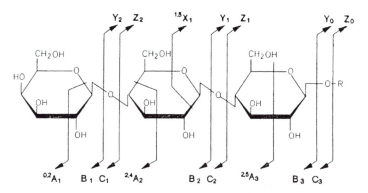

Figure 7.22 *Nomenclature to designate the fragments ions formed by the dissociation of oligosaccharides in a mass spectrometer*
(Source: J. Vath and C.E. Costello, in *Methods in Enzymology*, McCloskey (ed), Academic Press, New York, 1990 Vol. 193, Ch. 40, p. 743, Figure 2)

As for peptides and oligonucleotides, the sequence and structure of oligosaccharides can be assembled from the *m/z* values of their fragment ions. Appendix 10 shows the mass increments for common monosaccharide units, representing the molecular weight of a monosaccharide less 18 *u* for a molecule of water. To illustrate this process, the tandem mass spectrum of a pentasaccharide recorded in the negative ion mode is shown in Figure 7.23.

Fragment ions that result from glycosidic bond cleavages are evident in this spectrum in addition to those formed by ring cleavage (A and X ions). The appearance of abundant $^{2,4}A_3$ and $Y_{3\alpha}$ ions at *m/z* 545 and 586 enables the branched nature of the structure to be deduced.

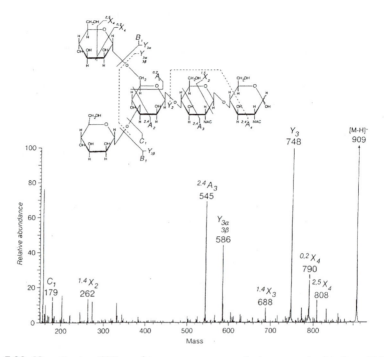

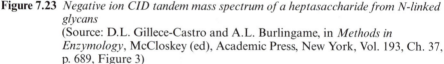

Figure 7.23 *Negative ion CID tandem mass spectrum of a heptasaccharide from N-linked glycans*
(Source: D.L. Gillece-Castro and A.L. Burlingame, in *Methods in Enzymology*, McCloskey (ed), Academic Press, New York, Vol. 193, Ch. 37, p. 689, Figure 3)

7.4.2 Exoglycosidase Digestion

An alternate strategy to tandem mass spectrometry is to digest the oligo-saccharide with a series of exoglycosidases and record the molecular weights of the resulting products. A number of commercial exogly-cosidases are available that show a range of specificities for oligo-saccharide cleavage. Sialidases, galactosidases and mannosidases can be employed to cleave glyosidic bonds at specific monosaccharides thus releasing these components. Figure 7.24 represents in cartoon form how glycosidases can be used to distinguish between alternate biantennary and triantennary structures by MS analysis after stepwise digestion.

7.4.3 Derivatisation Approaches: Oxidative and Reductive Cleavage to Identify Branching

There are many occasions where it is desirable, even necessary, to derivatise an oligosaccharide prior to its analysis. This is conducted

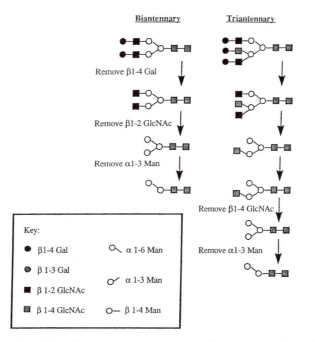

Figure 7.24 *Exoglycosidase digestion used to determine the structure of branched oligosaccharides*
(Source: adapted from R. Orlando and Y. Yang, in *Mass Spectrometry of Biological Materials*, 2nd edn, B.S. Larsen and C.N. McEwen (ed), Marcel Dekker, New York, USA, 1998, Ch.9, p. 236, Figure 11)

in order to improve the volatility of the molecule, increase the yield of parent ions, direct its fragmentation in a more predicatable manner or assist with the investigation of particular structural features. Methylation and acetylation reactions are among the most common derivatisation reactions employed. These reactions, however, show little specificity for functional groups within the sugar and as such are generally used to completely derivatise all such groups (known as permethylation and peracetylation) to improve the molecule's ionisation efficiency in the positive ion mode.

Periodate oxidation provides a more selective strategy for probing structural elements within an oligosaccharide. The reaction involves the cleavage of C-C bonds and allows for the branching pattern within O-linked oligosaccharides to be determined. This is illustrated in Figure 7.25 where a 1,4-linked and 1,6-linked hexose unit undergoes ring opening following oxidation and reduction to yield monosaccharide-units of different mass (208 and 164 u respectively) after methylation.

The same chemistries can be used to determine the anomeric con-figuration of glycosidic bonds by mass spectrometry. It is possible to

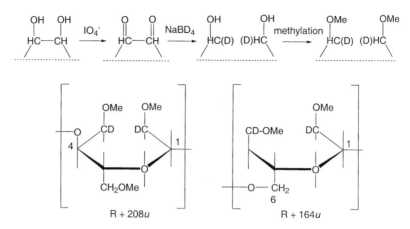

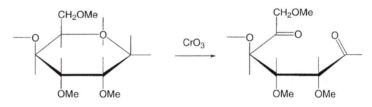

Figure 7.25 *Identification of 1,4-linked and 1,6-linked hexose after ring opening by periodate oxidation, reduction and methylation*

selectively oxidise the β-anomer of hexose, over its α counterpart, leading to a mass shift being detected in this region of the oligosaccharide (Figure 7.26).

Figure 7.26 *Oxidisation of the β-anomer of hexose, that occurs selectively over its α counterpart, can be identified by a mass shift*

FURTHER READING

G. Siuzdak, *Mass Spectrometry for Biotechnology*, Academic Press, 1996.

K.M. Downard Advances in protein analysis and sequencing by mass spectrometry, *New Adv. Anal. Chem.*, 2000, **P2**, 1–30.

A.G. Marshall, C.L. Hendrickson and S.D.H. Shi, Scaling MS Plateaus with FTICR MS, *Anal. Chem.*, 2002, **74**, 252A–259A.

W.J. Henzel, T.M. Billeci, J.T. Stults, S.C. Wong, C. Grimley and C. Watanabe, Identifying proteins from two-dimensional gels by molecular mass searching of peptide fragments in protein sequence databases, *Proc. Natl. Acad. Sci. USA*, 1993, **90**, 5011–5015.

J.R. Yates III, S. Speicher, P.R. Griffin and T. Hunkapiller, Peptide mass maps: A highly informative approach to protein identification, *Anal. Biochem.*, 1993, **214**, 397.

M. Mann, P. Hojrup and P. Roepstorff, Use of mass spectrometric molecular weight information to identify proteins in sequence databases, *Biol. Mass Spectrom.*, 1993, **22**, 338–345.

K. Biemann and I.A. Papayannopoulos, Amino acid sequencing of proteins, *Acc. Chem. Res.*, 1994, **27(11)**, 370–378.

I.A. Papayannopoulos, The interpretation of collision-induced dissociation tandem mass spectra of peptides, *Mass Spectrom. Rev*, 1995, **14(1)**, 49–73.

J.R. Engen, D.L. Smith, Analysis of proteins with hydrogen exchange and mass spectrometry, *Anal. Chem.*, 2001, **73**, 256A-265A.

S.D. Maleknia, K.M. Downard, Radical approaches to probe protein structure, folding, and interactions by mass spectrometry, *Mass Spectrom. Rev.*, 2001, **20(6)**, 388–401.

J.A. Loo, Studying noncovalent protein complexes by electrospray ionization mass spectrometry, *Mass Spectrom. Rev.*, 1997, **16(1)**, 1–23.

J.G. Kiselar and K.M. Downard, Antigenic Surveillance of the Influenza Virus by Mass Spectrometry, *Biochemistry*, 1999, **38(43)**, 14185–14191.

M.R. Wilkins, K.L. Williams, R.D. Appel and D.F. Hochstrasser, *Proteome Research: New Frontiers in Functional Genomics*, Springer Verlag, Berlin, Germany, 1997.

J.A. McCloskey, A.B. Whitehill, J. Rozenski, F. Qiu and P.F. Crain, New techniques for the rapid characterization of oligonucleotides by mass spectrometry, *Nucleosides & Nucleotides*, 1999, **18**, 1549–1553.

B. Reinhold, V. Reinhold and C.E. Costello, Carbohydrate molecular weight profiling, sequence, linkage and branching data: ESI-MS and CID, *Anal. Chem.*, 1995, **67**, 1772–1784.

K.H. Khoo and A. Dell, Assignment of anomeric configurations of pyranose sugars in oligosaccharides using a sensitive FAB-MS strategy, *Glycobiology*, 1990, **1** 83–911.

Mass Spectrometry in Medical Research

8.1 CHARACTERISATION AND QUANTITATION OF DRUGS AND METABOLITES

8.1.1 Introduction

In recent years, there has been an appreciable shift in focus in drug discovery away from the design, synthesis, characterisation and evaluation of organic drug molecules. Today's pharmaceutical and academic research laboratories seek to gain a global understanding of the genetic basis of disease states (through functional genomics), potential causative proteins or markers (functional proteomics), and the evaluation of drug candidates and therapies. Pharmacological studies of drug absorption, excretion and metabolism are also performed in the context of a complete description of human biology (metabolomics). The identification and characterisation of these biomarkers or targets can be performed by mass spectrometry using approaches described in Chapter 7.

The traditional characterisation and evaluation of organic drug targets, nonetheless, continues to be an important area of medical research. In these studies, molecules are designed to "target" a biological molecule, tissue or system to stimulate, confer or prevent some function or activity. Such molecules are usually constructed by synthetic routes either as a single lead compound or as part of a chemical library of related compounds. Potential drug targets are assessed based upon their bioavailability (or the extent to which the administered dose reaches its target), half-life, and therapeutic index. The half-life of the compound represents the time in which it takes for 50% of the concentration of the compound to be excreted or metabolised *in vivo*. The therapeutic index represents a measure of the desired function or activity versus any undesired side effects. The most effective drugs are those with a high therapeutic index, a high bioavailability, and often a low half-life.

8.1.2 Sample Preparation Techniques in Drug Discovery

When dealing with biological and *in vitro* samples, the preparation of the sample for mass spectrometric analysis is of paramount importance in order to achieve analytical success. A wide variety of approaches including precipitation and centrifugation methods, ultrafiltration, solid-phase extraction, blotting and immobilisation are all employed subject to the sample at hand and its biological source. The precipitation of biological components from an extract can be achieved by adding denaturing solvents such as methanol, acetonitrile or acetone. The recovery of these components follows the rapid mixing of the solution and their density centrifugation to the base of the sample vial.

Ultrafiltration provides another means with which to separate the components of biological mixtures by passing the solution through molecular weight cut-off filters. Under centrifugation the larger molecular weight species remain trapped on the top of the filter while smaller compounds pass through with the solvent. This process can be performed in a series of stages so that a sample mixture is effectively partitioned into sets of compounds spanning several molecular weight ranges.

Solid-phase extraction (SPE) takes advantage of the separation characteristics exploited in liquid chromatography. Samples are loaded onto small cartridges, pre-packed with chromatographic supports suitable for reverse-phase, ion exchange or ion exclusion chromatography. The sample solution is passed through the cartridge under gravity, by vacuum or using centrifugal force. Solvents of differing compositions are then added in succession to effect the partitioning of solutes between the solid and solution phases.

Blotting and immobilisation procedures are also used widely in drug discovery applications to isolate particular components from biological sources. Drug targets can also be absorbed or immobilised on films and screened against an array of drug compounds ahead of their analysis. The screening of drugs using mass spectrometry is the subject of Section 8.4.

8.1.3 Qualitative Analysis of Organic Drugs and their Metabolites

The identification or characterisation of a drug compound by mass spectrometry is dependent upon the structural features of that compound. For most small to moderately sized (~1000 Da.) compounds this is achieved by either GC-MS or LC-ESI-MS. The former approach pre-dated the development of the ESI technique and has to some extent been superseded by it. However, certain volatile or derivatised molecules are particularly suited to GC-MS in which the resulting EI (or CI) mass spectrum provides both molecular weight and structural information.

The latter is usually achieved through the use of tandem mass spectrometry in LC-ESI-MS experiments since the "soft" nature of the ESI technique results in most compounds resisting fragmentation. Such approaches can be used for the characterisation of trial drugs, those approved for pharmaceutical use, or illegal drugs or narcotics.

As an illustration, cocaine and 6-acetylmorphine can be detected in a single human hair by GC-MS. A cutting from a single hair was washed and heated in methanol. The volatile components were passed into the ion source of a GC-MS and both cocaine and 6-acetylmorphine were detected by means of selected ion monitoring (SIM) (Chapter 5, Section 5.4.2) of the fragment ions at *m/z* 182 (Figure 8.1) and 268 respectively.

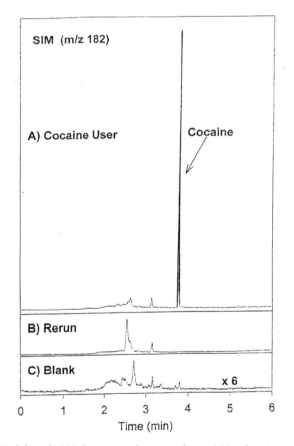

Figure 8.1 *SIM of the m/z 182 fragment of cocaine from: (A) a 1-cm piece of a cocaine user's hair (subject 2), (B) The rerun of the same hair sample used in (A) demonstrating the near 100% recovery, and (C) a 1-cm piece of hair obtained from a drug free individual*
(Source: S.B. Wainhaus, N. Tzanani, S. Dagan, M.L. Miller and A. Amirav, Fast Analysis of Drugs in a Single Hair, *J. Am. Soc. Mass Spectrom.*, 1998, **9**, 1311, Figure 4)

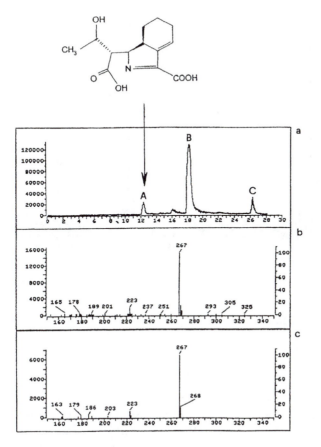

Figure 8.2 *Particle beam/chemical ionisation data for: (a) total ion current (m/z 700–1300) chromatogram of rat urine, (b) mass spectrum of component "A" the metabolite eluting at 12.6 min., and (c) mass spectrum of metabolite reference standard*

(Source: adapted from L. Iavarone, M. Scandola, F. Pugnaghi and P. Grossi, Qualitative Analysis of potential metabolites and degradation products of a new antiinfective drug in rat urine, *J. Pharma. Biomed. Anal.*, 1995, **13**, 607–614, Figure 8)

These constituents were detected at concentrations as low as 10 parts-per-billion (ppb) with analyses conducted within 10 minutes.

The qualitative analysis of metabolites is also an important requirement of drug testing to evaluate both the half-life of a drug and the nature and potential toxicity of its metabolites.

LC-MS has been employed to follow the metabolic fate of a β-lactam antibiotic. The metabolic profiles of the antibiotic in rat urine were monitored by mass spectrometry in conjunction with ion exchange HPLC after a single intravenous administration of the drug. Two

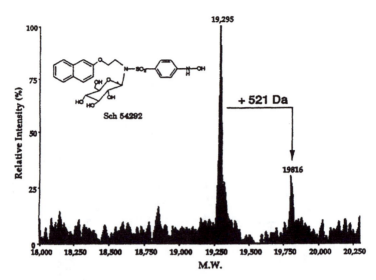

Figure 8.3 *Deconvoluted (molecular weight adjusted) ESI mass spectrum of a 1:1 ras GDP oncoprotein inhibitor (SCH 54292) complex with an average mass of 19,816 Da* (Source: A.K. Ganguly, B.N. Pramanik, E.C. Huang, S. Liberles, L. Heimark, *et al.*, Detection and structural characterisation of Ras oncoprotein-inhibitors complexes by electrospray mass spectrometry, *Bioorganic and Medicinal Chemistry*, 1997, **5(5)**, 817–820, Figure 1)

metabolites were detected, one of which corresponds to a ring-opened degradation product based upon its mass spectrum (Figure 8.2).

The association of drugs with protein targets can also be evaluated by mass spectrometry. As described in Chapter 7 (Section 7.5.2), ESI and to a lesser extent MALDI mass spectrometry are able to both preserve and detect specific solution state associations within a mass spectrometer. A 1:1 non-covalent complex between a ras GDP oncoprotein and a potential inhibitor is shown in Figure 8.3. Mutant ras proteins have been implicated in the growth of a wide range of human tumours and thus the inhibition of GDP could prevent continued tumour growth. The approach offers the opportunity to study protein complexes that cannot be investigated by other methods. In this case, crystals for the ras GDP protein could not be successfully prepared for X-ray crystallography.

The purity of a drug compound must also be thoroughly evaluated before it can be used in clinical trials. In some cases it is necessary that the drug be enantiomerically pure. Where a chiral drug has a particular potency and the drug is to be administered as a racemic mixture, it is necessary to establish that the inactive enantiomer affords no potential side effects.

The kinetic method developed by Cooks (Section 6.1.4) has been used to determine the enantiomeric composition of a drug mixture through the competitive dissociation of their copper-bound ion complexes. Copper(II)-bound ion complexes formed from seven model drugs together with a series of chiral reference compounds (L–amino acids only) were analysed by electrospray ionisation mass spectrometry. The complexes were found to undergo collisionally activated dissociation (CAD) by competitive loss of either the neutral drug molecule or the reference. The ratio of the two competitive dissociation rates allowed the composition of enantiomeric drugs in the mixture to be determined using a two-point calibration curve.

8.1.4 Quantitative Analysis of Drug Compounds and their Metabolites

Mass spectrometry plays a central role not just in confirming the presence or structure of a drug molecule, but also in measuring the absolute and relative levels of the compound or its metabolites in serum or plasma. A measurement of the concentration of the compound or its metabolites as a function of time, after the dose has been administered, provides a pharmacokinetic profile that is useful in establishing a drugs' bioavailability or its rate of metabolism throughout the body. Quantitation measurements are also necessary where the metabolite is toxic or pharmacologically active when it reaches a particular concentration.

In addition to selected ion monitoring (SIM) described above, *selected* or *multiple reaction monitoring* (SRM or MRM) is one of the most common approaches used for this purpose and is achieved within a tandem mass spectrometer. Here the metastable transition or conversion of a metabolic precursor to a product is monitored. Only selected precursor ions that decompose to product ions of a particular *m/z* ratio will be detected. This affords optimal sensitivities where compounds can be quantitated to the sub-ppb (part-per-billion) level. In a MRM experiment, a series of such reactions is monitored by rapid switching of the electric fields applied to the mass analyser to study each reaction in turn.

Figure 8.4 shows the MRM ion chromatograms for seven dosed compounds extracted from brain tissue plus an internal standard (top chromatogram).

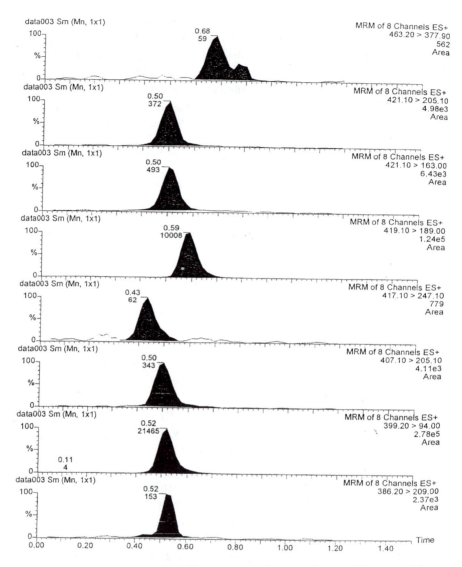

Figure 8.4 *ESI positive ion MRM chromatograms for seven compounds extracted from brain tissue plus an internal standard (top chromatogram)*
(Source: D.T. Rossi, Sample preparation and handling for LC/MS in drug discovery, in *Mass Spectrometry in Drug Discovery*, D.T. Rossi and M.W. Sinz (ed), Marcel Dekker, New York, 2002, Ch. 6, p. 206, Fig. 15)

8.2 DEFINING METABOLIC PATHWAYS WITH MASS SPECTROMETRY

Where the pharmacokinetic profile of a metabolite exists, it is often desirable to establish the reaction pathway through which the metabolite is produced. Common metabolic processes include oxidation, methylation, acetylation and epoxidation as well as degradation reactions. These consequently cause changes in the molecular weight of the product in each step of a metabolic pathway.

Heavy isotopes (*e.g.* ^{13}C, ^{15}N) are of use for tracking the decomposition or reaction of a drug as it is metabolised. The isotopic enrichment of a drug changes its molecular weight and isotopic profile (Chapter 1, Section 1.2). If the heavy isotope is retained by the metabolite its ion will have a *m/z* ratio and isotopic distribution that will differ in appearance from that of its unlabelled form. Thus the drug and metabolite can be associated. The use of a series of labels at different positions throughout the drug enables a range of metabolites to be identified from which their dissociation pathways can be deduced. The placement of the heavy isotope should be decided upon with some care and it must be placed in a metabolically inert position since if the isotope "lost" early during its metabolism, the reaction pathway can no longer be followed.

8.3 CHARACTERISATION OF DRUG LIBRARIES BY MASS SPECTROMETRY

As mentioned at the introduction to this chapter, drugs and drug targets are usually constructed as part of a chemical library. These libraries are pooled and split and subsequent sets of compounds screened for activity in an assay. Library components can be characterised in these sets by reverse phase HPLC or capillary electrophoresis coupled to an ESI or APCI mass spectrometer (Chapter 3, Section 3.2.10). It is common to pass a portion of the eluant from the column to a secondary detector, such as an ultraviolet (UV) absorbance detector preferably operating over a range of wavelengths. The purpose of the secondary detector is to assist with the detection and quantitation of the components.

Stable isotopes can also be used to encode particular components of drug libraries to assist with their identification in what is termed *stable isotope encoding*. In many instances, libraries are constructed of structurally similar compounds whose molecular weights may coincidentally be identical or indistinguishable by mass spectrometry. If stable heavy isotopes are incorporated into some compounds during their synthesis, these can be mixed in series at varying molar ratios with their non-labelled counterparts, such that each component can be distinguished based upon the ratio of the ion signals for the labelled and unlabelled forms.

Establishing the chemical components in each library enables screening of the activity of these components to begin. For the most part, this is performed without the aid of a mass spectrometer but a number of mass spectrometric-based assays have now been developed that are of use in drug discovery investigations.

8.4 DRUG SCREENING USING MASS SPECTROMETRY

The immobilisation of a drug target to a surface or membrane provides a means with which to screen drug libraries in an automated manner. A solution of mixtures of library compounds can be passed across the bound target and the surface washed after an appropriate incubation time. The bound drugs can then be chemically released and detected.

Mass spectrometry has been employed in such assays to detect the released drug including the use of MALDI to catalyse that release and subsequently ionise the bound drug. A multi-sample MALDI target or miniaturised chip format enables a series of drug interactions to be studied simultaneously where each position on the target or chip characterises a unique association. Such an approach has been used to screen protein associations in what has been described as *biomolecular interaction analysis mass spectrometry* (BIA-MS) (Figure 8.5).

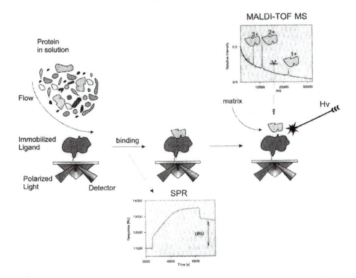

Figure 8.5 *Schematic representation of the BIA-MS method in which surface-immobilised ligands with affinities toward a protein of interest are used to retrieve the protein from a complex biological mixture. Surface plasma resonance is used to monitor the interaction and quantify the amount of retrieved protein(s). MALDI-MS releases the bound protein enabling it to be identified by molecular weight* (Source: D. Nedelkov and R.W. Nelson, Biomolecular interaction analysis mass spectrometry: A comprehensive microscale proteomics approach, *American Laboratory*, 2001, 22–25, Figure 1)

In an extension of this approach, Williams, Nelson and co-workers have used a 96-well robotic workstation to preferentially isolate proteins in parallel from human blood and plasma using tips filled with immobilised antibody to each protein target. The isolated proteins were subsequently transferred to a MALDI sample target for their detection by mass spectrometry (Figure 8.6). The affinity isolation of proteins from such biological matrices in combination with MALDI-TOF MS holds promise in proteomics for the detection of protein markers or mutants associated with genetic abnormalities and disease.

8.5 TRACE ELEMENT ANALYSIS IN NUTRITION

Just as isotopic labelling techniques are used to follow the absorption and metabolism of a man-made drug administered orally or intra-venously, mass spectrometry can also be applied to follow the ingestion and absorption of essential vitamins and minerals. Because not all nutrients and minerals in a diet are retained and available for physio-logical function, the uptake of dietary elements can be followed using stable heavy isotopes.

Dietary iron is vital for correct physiological function and a lack of iron leads to anemia, impaired mental and motor development, and a reduced resistance to infection. The absorption of iron by red blood cells can be followed using test meals isotopically-enriched with ^{57}Fe or ^{58}Fe. Subjects are fed meals containing either ^{57}Fe or ^{58}Fe and the amount of isotope retained in the blood is quantified by mass spectrometry. *Inductively coupled plasma mass spectrometry* (ICP-MS) is used widely for such an analysis (see Section 9.1.1) and employs a high resolution mass analyser and multiple ion detectors.

ICP-MS has also been employed to study calcium uptake in bone. During bone growth, calcium is transferred into bone at rates greater than is lost. The balance alters with age resulting in a reduction in bone mineral density and the onset of osteoperosis. Women are at particular risk of osteoperosis and the resultant bone fractures. Stable isotope studies using diets containing ^{40}Ca and ^{41}Ca allow the absorption of calcium to be followed by monitoring calcium levels excreted in fecal matter or urine, or present in blood plasma.

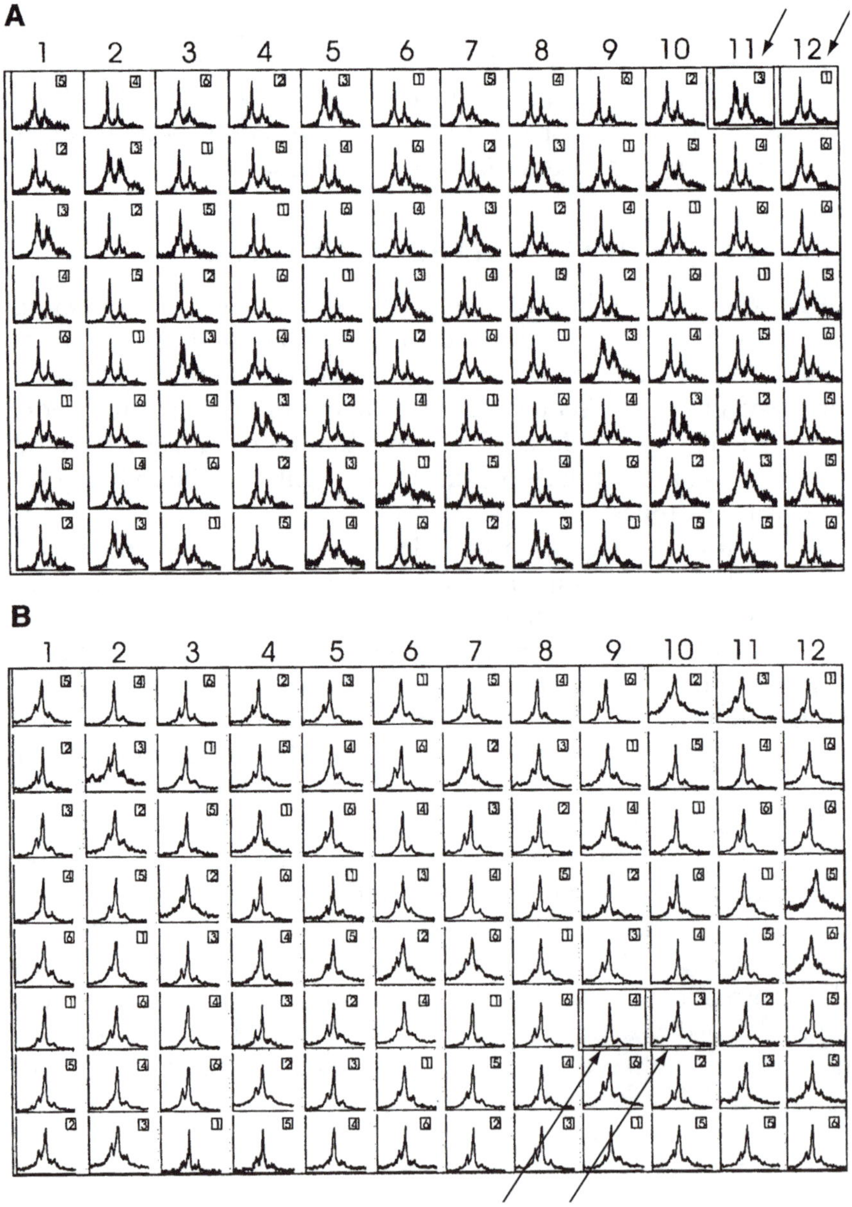

Figure 8.6

FURTHER READING

D.T. Rossi and M.W. Sinz (ed) *Mass Spectrometry in Drug Discovery*, Marcel Dekker, New York, 2002.

D.I. Papac and Z. Shahrokh, Mass spectrometry innovations in drug discovery and development, *Pharmaceutical Research*, 2001, **18(2)**, 131–145.

S.J. Gaskell and D.S. Millington, Selected metastable peak monitoring: A new specific technique in quantitative gas chromatography mass spectrometry, *Biomed. Mass Spectrom.*, 1978, **5**, 557–558.

U.A. Kiernan, K.A. Tubbs, K. Gruber, D. Nedelkov, E.E. Niederkofler, P. Williams and R.W. Nelson, High-throughput protein characterization using a mass spectrometric immunoassay, *Anal. Biochem.* 2002, **301**, 49–56.

F. Mellon, R. Self and J.R. Sartin, *Mass Spectrometry of Natural Substances in Food*, Royal Society of Chemistry, Cambridge, UK.

Figure 8.6 *(opposite) High-throughput Mass Spectrometric Immunoassay (MSIA) analysis of transthyretin (TTR) and retinol binding protein (RBP) using human plasma samples from six individuals randomly arranged in a 96-well titer plate. (A) Mass spectra result from MSIA analysis utilising anti-TTR derivatised pipette tips. Shown is the region of the singly charged TTR signals. Highlighted cells show spectra with resolvable parent ion differences between each other. (B) Mass spectra result from the MSIA analysis of the same samples utilising anti-RBP derivatised pipette tips. Shown is the region of the doubly charged RBP signals*
(Source: U.A. Kiernan, K.A. Tubbs, K. Gruber, D. Nedelkov, E.E. Niederkofler, P. Williams and R.W. Nelson, High-Throughput Protein Characterization Using Mass Spectrometric Immunoassay, *Anal. Biochem.*, 2002, **301**, 49–56, Figure 3)

CHAPTER 9

Mass Spectrometry in the Environmental and Surface Sciences

9.1 ENVIRONMENTAL ANALYSIS

The mass spectrometer is the most widely used detector to analyse the effects of natural and man-made substances on the environment. Social pressures as well as the need for ecological sustainability have necessitated that changes to our environment are monitored with some precision. The Earth's growing population has placed increased demands on the planet's resources and there is now a greater appreciation than ever before of the affect of man-made pollutants and wastes on the environment and their impact on climate change. The compounds studied by mass spectrometry include heavy metals, man-made pesticides, components of industrial waste and their by-products, disinfecting agents, explosives, and those excreted or obtained from naturally occurring algae, toxins and microorganisms. These substances are often present at trace levels within complex mixtures and can only be studied using a mass spectrometer.

9.1.1 Heavy Metals and Elemental Analysis

Although trace concentrations of some metals such as iron and zinc obtained from certain foods are essential to our well-being, others such as lead and mercury have long been known to cause serious adverse effects to human health. Exposure to lead in paints, petrol and other industrial waste, for example, has long been associated with mental illness and even paralysis. Lead has four natural isotopes ^{204}Pb, ^{206}Pb, ^{207}Pb and ^{208}Pb of which the latter three are produced by radioactive decay of isotopes of uranium and thallium.

9.1.1.1 Thermal Ionisation Mass Spectrometry. Thermal ionisation mass spectrometry (TIMS) is a common technique applied to the analysis of elements at low ng g^{-1} levels in environmental samples. In TIMS, a small volume of an aqueous sample solution containing between µg and ng of the element of interest is deposited onto a clean filament surface and evaporated to dryness. The filament, usually composed of a thin film of ruthenium, is heated to thermally evaporate the sample. A second like-filament that emits electrons to ionise the sample. Thermal ionisation is highly selective, with different elements ionised according to the filament temperature at any point in time. Isotopes are separated and detected simultaneously using a magnetic sector mass spectrometer equipped with a multiple ion collector.

9.1.1.2 Inductively Coupled Plasma Mass Spectrometry. Inductively coupled plasma mass spectrometry (ICP-MS) is also widely used for elemental analysis (Figure 9.1). The sample, usually in a liquid form, is introduced at approximately 1 ml min^{-1} via a nebuliser where it is converted into a fine aerosol with a gas (normally argon). The fine droplets of the aerosol are transported into the plasma torch *via* a sample injector where they are converted from a liquid aerosol to a solid and then to a gas. This sample gas is then "atomised" and ionised. Ionisation is achieved by electron impact using electrons from a high-voltage spark. This results in a high temperature (~7,000 K) ion plasma being emitted from the open end of the tube which reflects the elemental composition of the sample. Solid samples can be analysed directly by *laser ablation ICP-MS*. Here a portion of the sample is ablated by a laser in an inert atmosphere (usually of argon) under atmospheric pressure and then atomised and ionised in the plasma.

Most commercial ICP mass spectrometers contain a single ion detector to detect elements to low part-per-trillion (ppt) levels. However, specialised magnetic sector ICP-MS instrumentation fitted with multiple detectors are also used for isotope ratio analysis (see also Section 9.2).

The concentrations of heavy metals such as chromium (Cr) in soil and

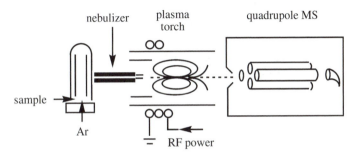

Figure 9.1 *Schematic representation of an inductively coupled plasma mass spectrometer (ICP-MS) featuring a quadrupole mass analyser*

water samples have been analysed by ICP-MS. Cr(VI)-containing compounds are found in the environment as a result of industrial pollution and fertilisers. Ion exchange chromatography has been coupled to ICP-MS to measure chromium levels in waters from industrial waste and sewage treatment plants. Both cationic Cr(III) and anionic Cr(VI) species were detected below 0.5 µg/1 by monitoring [52]Cr using ICP-MS.

A disadvantage of ICP-MS is that there is a multitude of plasma-derived chemical reactions that can occur, resulting in molecular ions that can mask or diminish ions of the elements of interest. In order to improve detection limits and/or reduce the amount of sample that needs to be analysed, alternate low-pressure helium sources have been constructed with collision cells used to dissociate molecular ions formed within the plasma.

9.1.1.3 Isotope Dilution. Isotope dilution is a method used in conjunction with TIMS and ICP-MS to quantitate the elemental abundances of trace elements that contain two or more naturally occurring isotopes. It does so with unmatched sensitivity and accuracy and thus is in widespread use. The isotope dilution method involves spiking and blending the sample (S) of interest with a reference (R) material containing the same element at known concentration. The quantitation of trace elements in the samples is then determined by equation 9.1 where Q_S is the element concentration of interest, Q_R is the concentration of the element in the reference material, R the ratio of isotopes in the reference, S the isotope ratio of the same isotopes in the sample, B the isotope ratio of the elements in the blend, m_S the atomic mass of the lighter isotope for the element, and m_R the atomic mass of the heavy isotope.

$$Q_s = Q_R(R - B)\, m_S/(B - S)m_R \qquad (9.1)$$

To minimise errors, the purities of samples S and R must be approximately the same and the mass spectrometric measurements recorded under the same conditions. When this is observed, measurements to within 0.1% of true values are possible.

TIMS in conjunction with the isotope dilution method has been used in environmental research to determine heavy metals in the atmosphere. The presence of thallium in the atmosphere over Antarctica was measured at 0.2 pg m^{-3} using this approach.

9.1.2 Organic Pesticides

GC and LC coupled mass spectrometric-based methods are preferred over other analytical approaches for the characterisation and quantitation of organic pesticides. New pesticides are developed every year, in part to keep abreast of the resistance of insects and other pests to existing ones. For each new pesticide, a thorough evaluation of its fate in the environment is required including knowledge of its transport properties and degradation products. Pesticides can accumulate in soils, infect ground water and water supplies and be ingested by humans by way of crops and livestock. An accurate measure of the levels of pesticides in the environment is routinely needed in order to minimise their adverse effects on human health.

Some modern pesticides are active as a single enantiomeric form such that regulatory authorities mandate that only this enantiomer may be administered where the inactive enantiomer has a low degradation rate. This necessitates that enantiomeric forms of a pesticide in a racemic mixture are resolved and independently quantified in environmental samples.

As an illustration, soil samples from the south-east of Spain treated with the propionic acid-derived herbicides mecoprop and dichloroprop were characterised by HPLC using a chiral stationary phase and GC-MS. Selected ion chromatograms from the GC-MS analysis of mecoprop (MCPP) (*m/z* 169) and dichlorprop (DCPP) (*m/z* 189) show that the methylated forms of *R,S*-MCPP and *R,S*-DCPP are clearly resolved in two soil samples (denoted a and b in Figure 9.2).

Quantitation of the ion signals demonstrated detection levels of greater than 80% for both MCPP and DCPP from control soil-doped experiments. For each pesticide, the *R*-enantiomer was found to degrade faster than its inactive *S*-form.

9.2 ISOTOPE RATIO MASS SPECTROMETRY

Beyond the characterisation and quantitation of compounds in the environment, isotope ratio mass spectrometry can be used to distinguish the source of such chemicals. The uptake of water, nitrogen and carbon in plants, nutrient transfer in aquatic ecosystems, and the rates of chemical and biochemical degradation processes can all be studied by mass spectrometry.

Isotope Ratio Mass Spectrometry (IRMS) is performed using a specially constructed mass spectrometer designed to maximise ion beam stability and sensitivity at the expense of mass resolution. These instruments (Figure 9.3) usually feature an electron impact source with a gas inlet. As in a single magnetic sector mass spectrometer (described in Section 3.3.2), ions are accelerated down a flight tube between a magnet where their curved trajectories depend on the ion's *m/z* ratio (equation

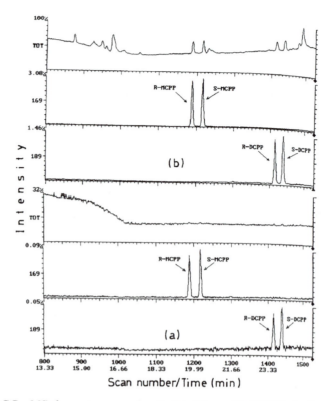

Figure 9.2 *GC – MS chromatograms of methylated* R- *and* S-*MCPP plus* R- *and* S-*DCPP*
in a silty loam sample (a) without and (b) with added peat. In both cases, the
total-ion chromatogram and the single-ion chromatograms corresponding to
m/z *values of 169 (specific for the enantiomers of MCPP) and 189 (for those*
of DCPP) are shown
(Source: F. Sánchez-Rasero, M.B. Matallo, G. Dios, E. Romero and A. Peña,
Simultaneous determination and enantiomeric resolution of mecoprop and
dichlorprop in soil samples by high-performance liquid chromatography
and gas chromatography – mass spectrometry, *J. Chromatography A*, 1998,
799(1–2), 355–360, Figure 2)

3.22). Instead of scanning the magnetic field or accelerating voltage,
these values are fixed for a particular measurement to transmit to the
Faraday cup detector (Section 3.4.1) ions with a *m/z* range of just a
few mass units. The mass limit of an isotope ratio mass spectrometer is
typically 100.

Slits on the instrument are large to maximise ion transmission. As a
result, the ion signals typically have a flat top (Figure 9.4). Isotope
ratios are measured based upon the area under the isotopic ion
signals. The analytes are often simple gases such as H_2, O_2, N_2 and CO_2
or pyrolysis or combustion products formed by heating of the sample at

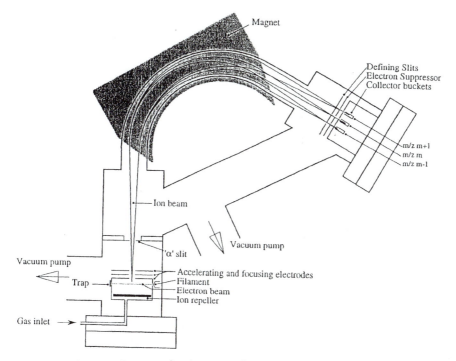

Figure 9.3 *Schematic diagram of an isotope ratio mass spectrometer*
(Source: A. Barrie and S.J. Prosser, in *Mass Spectrometry of Soils*,
T.W. Boutton and S.-I. Yamasaki (ed), Marcel Dekker, New York, 1996,
Ch. 1, p. 9, Figure 1)

high temperatures. A gas chromatograph can be used to separate these
combusted species prior to entry into the ion source.

9.3 PORTABLE MASS SPECTROMETERS

There are many instances where it is desirable to analyse environmental
compounds or organisms in the field. A variety of miniature, portable
mass spectrometers have been used for this purpose. Small quadrupole
ion trap and time-of-flight mass analysers are typically employed
over heavier magnetic-based instruments. These truck or hand-portable
mass spectrometers are used by government laboratories, academic
researchers and by the military in applications as diverse as environmental monitoring, forensic science, oceanographic research, and the
detection of nerve gas and other agents of warfare.

Field portable GC-MS mass spectrometers feature a sensing device or
sniffer with which to sample volatile agents in air. Submergeable mass
spectrometers have also been developed for oceanographic research to

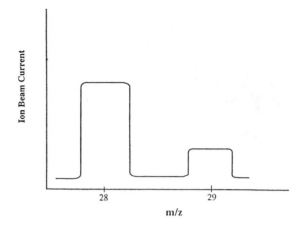

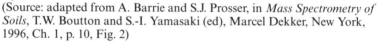

Figure 9.4 *Ion signals measured on a isotope ratio mass spectrometer typically have a flat top*
(Source: adapted from A. Barrie and S.J. Prosser, in *Mass Spectrometry of Soils*, T.W. Boutton and S.-I. Yamasaki (ed), Marcel Dekker, New York, 1996, Ch. 1, p. 10, Fig. 2)

monitor water quality, including the influences of tidal flows or the effects of dredging on water quality. These analytical measurements have particular challenges for a mass spectrometer including maintaining a vacuum in such an instrument underwater, providing a sufficient power source during the course of the experiments, transmitting the data to the shore, and sampling the analytes underwater. This latter challenge can be met by using *membrane introduction mass spectrometry* (MIMS) in which a sample is exposed to a semi-permeable membrane through which compounds are selectively transferred into the mass spectrometer. Components, sampled from such environments, can be analysed in concentrations down to the parts-per-trillion level.

Despite their ban under the Geneva Protocol of 1925, chemical warfare agents have continued to be produced and in some cases inflicted on both military and civilian populations. Military personnel are particularly susceptible to exposure to chemical weapons. These include mustard gas and lewisite that cause blistering of the skin, diphenylcyanoarsine that leads to vomiting, tearing agents and the nerve agents sarin, tabun and so-called VX. Aircraft, ground-based vehicles or personnel can carry portable mass spectrometers to check for such agents prior to the deployment of large numbers of troops. The EI mass spectrum of mustard gas or 1,1-thiobis(2-chloroethane) is shown in Figure 9.5. Mustard gas is a highly fat soluble compound and accumulates in tissues with a high fat content. Absorption of just a few milligrams of mustard gas within human tissue can be fatal.

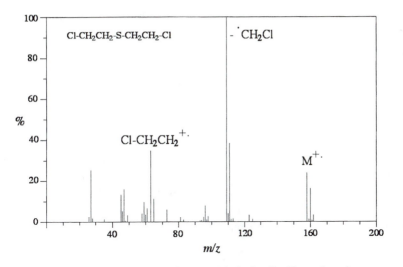

Figure 9.5 *EI mass spectrum of mustard gas or 1,1-thiobis(2-chloroethane)*
(Source: RSC database)

9.4 CHEMISTRY OF THE EARTH'S IONOSPHERE

The Earth's atmosphere and ionosphere are rich in ion chemistry made possible due to the production of ions by electrical discharges and radiation. A significant understanding of the chemistries above the surface of the planet has been achieved by experiments that mimic such processes in the laboratory and by *in situ* measurements using rocket-borne mass spectrometers. Of particular interest is the effect of man-made pollutants on the Earth's atmosphere.

The Earth's atmosphere is divided up by altitude according to a range of criteria where the ionosphere is described as the region where free electrons exist at some 60–1000 km above the Earth's surface. At the highest altitudes, levels of solar radiation are the greatest but since there are few atoms or molecules present, few ions form. Below the ionosphere, electrons spontaneously react with gases to form both positive and negative ions. From the standpoint of ion chemistry, most can be classified as occurring at low altitudes below some 80 km, or at high altitudes or low pressures above this height. The lower atmosphere above 10–15 km is defined as the stratosphere and below that the troposphere where the major gases present are oxygen (21%) and nitrogen (78%) together with argon and varying amounts of water, carbon monoxide and dioxide, and nitrogen dioxide.

In the upper atmosphere, solar radiation reaches the earth at a power of 1.37 kJ m^{-2} s^{-1}. The photoionisation of an atom of oxygen by solar radiation in the ionosphere results in $O^{+\cdot}$ (Chapter 2, equation 2.1) that can subsequently react to form the molecular ions $NO^{+\cdot}$ and O_2^+ (equations 9.2 and 9.3).

$$O^{+\cdot} + N_2 \rightarrow NO^{+\cdot} + N^\cdot \qquad (9.2)$$

$$O^{+\cdot} + O_2 \rightarrow O_2^+ + O^\cdot \qquad (9.3)$$

Rates for these reactions have been measured in the laboratory within high pressure and flowing afterglow mass spectrometers to be approximately 1.2×10^{-12} cm^3 s^{-1} and 2×10^{-11} cm^3 s^{-1} respectively at 300 K.

Cluster ions are also evident in the ionosphere formed by the weak association of water, nitrogen (equation 9.4) and carbon dioxide with ions in three-body associations.

$$NO^{+\cdot} + nH_2O + N_2 \rightleftharpoons NO^{+\cdot}(H_2O)_n(N_2) \qquad (9.4)$$

Cosmic rays are the primary ionisation source in the stratosphere and upper troposphere. One reaction of interest within the stratosphere is the recombination of molecular oxygen with oxygen atoms to form ozone. This gas reaches a peak density of a few parts per million at an altitude of 25 km above the Earth. The ozone layer protects the Earth from ultraviolet (UV) radiation that can penetrate the atmosphere by absorbing this radiation and dissociating back to its constituents. Negatively charged ions have been postulated to regulate ozone levels in the lower atmosphere as illustrated in equation 9.5 though this has been questioned based on a measured rate for this reaction.

$$NO_3^- + O_3 \rightarrow NO_2^- + 2O_2 \qquad (9.5)$$

It has further been shown that most ions do not react with man-made pollutants such as trichlorofluorocarbon ($CFCl_3$) and as such these pollutants cannot be depleted before they reach the stratosphere where they degrade ozone levels.

9.5 MASS SPECTROMETERS IN SPACE

It might be argued that the best environment in which to operate a mass spectrometer is one in natural vacuum. Beyond the environment within and around our own planet, mass spectrometers have been employed in space to study the cosmos. Expeditions to Mars, investigations of interstellar dust and the chemical composition of the tails of comets have all involved mass spectrometers. Earlier mass spectrometry experiments were performed on rocks returned to Earth during the Apollo missions.

9.5.1 Apollo Missions

Samples of the lunar surface returned to Earth as part of the six Apollo missions 11–17 (except the ill-fated Apollo 13 mission!) were subsequently examined by mass spectrometry. NASA scientists used *secondary ion mass spectrometry* (SIMS) (Section 3.2.5) to analyse the elemental composition of the moon rocks. This data has shown the moon to be enriched in the elements aluminium, uranium and thallium, in addition to ferric oxide, over that found on Earth. Samples obtained from the lunar highlands are rich in potassium, phosphorous and rare earth elements. The distribution of these elements on the lunar surface provides information about how the lunar crust was formed and has evolved over time.

Samples of finely pulverised lunar surface material were also analysed for traces of organic compounds by GC-MS. Samples were volatilised directly, after extraction with benzene and methanol, or extraction and pre-treatment with hydrogen fluoride and chloride. A number of organic compounds were detected at concentrations of less than 1 ppm, but all could have been associated with contamination on Earth and therefore no evidence of life (as we know it) on the surface of the moon has been found to date.

9.5.2 Viking and Mars Express Missions

To preclude the possibility of contamination of returned specimens, the objectives of the Viking experiments were to sample the composition of the atmosphere at the surface of Mars and to identify any volatile organic and inorganic compounds on the surface at the landing site. These data were measured directly at the surface of Mars and transmitted electronically back to Earth. The Viking spacecraft of the 1975 mission transported a Mars lander fitted with a gas chromatograph mass spectrometer designed for isotope ratio experiments.

Analysis of the atmosphere found the isotopic levels of oxygen to be within 10% of those values on Earth. A higher level of ^{15}N (74%), however, was detected at the surface of Mars with an isotope ratio of $^{15}N/^{14}N$ of 0.0064 (without background correction). ^{38}Ar was also detected above the Martian surface.

Soil samples collected at the Martian surface were robotically loaded into a pyrolysis source at the lander site and heated to 500 °C in a series of three chambers. The volatiles were passed into the carrier gas stream of a gas chromatograph interfaced to a mass spectrometer operating over a mass range of 12–200. The data, together with operating parameters, were transmitted back to Earth from the Martian surface. Only water and carbon dioxide were detected in these measurements and there was no evidence or organic compounds above the detectable levels of a few parts-per-billion. The absence of such compounds at the lander site does not preclude that life as we know it could be found elsewhere on the planet surface or interior. It is also possible that traces of micro-organisms in the samples did not release sufficient levels of organic material for detection or that such organic matter had been destroyed on the surface by solar radiation.

An opportunity to re-evaluate the planet's surface was afforded by the European Mars Express mission of late 2003 in a British-led expedition. The Beagle 2 lander carried an onboard minature 90° sector mass spectrometer with a magnet weighing less than 1 kg. It was fitted with a dual inlet source so that light elemental samples and standards can be sequentially studied in high precision isotope ratio measurements. Regrettably, signals from Beagle 2 were lost leaving its fate unknown and mysteries of the Red planet firmly intact.

9.5.3 Composition of a Comet

Unlike Mars, there is little doubt that organic compounds are ubiquitous within the nuclei of comets. There is considerable data to support the existence of molecular species within a comet. Data from ground-based spectroscopic measurements have been supplemented by both spectroscopic and mass spectrometric measurements in the Giotto and Vega missions to Halley's comet. Water makes up about 80% of the volatile content of the comet but hydrogen cyanide, carbon monoxide and dioxide, methanol, ammonia and formaldehyde have all been detected. Many of these molecules are likely to be ionised fragments of even larger parent molecules. The identity of these parent molecules, however, is not known since it is virtually impossible to observe the comet surface directly, even when a spacecraft is nearby.

The dust from Halley's comet was also examined using mass spectrometers on board both the Giotto and Vega probes. Of primary interest was the detection of intermediate-sized organic compounds that gave rise to mass spectra with ion signals separated by repeating 14–16 mass units. These ions indicated that the molecules exhibit a linear polymeric structure interpreted to be a signature of hydrocarbons, with a repeating $-(CH_2)_n-$ structure. The Giotto and Vega spacecraft further revealed a substantial enrichment of ^{12}C over that observed on Earth indicating an interstellar source for some of the organic compounds.

9.6 APPLICATIONS OF SECONDARY ION MASS SPECTROMETRY TO MATERIALS SCIENCE

Beyond moon rocks and interstellar particles, *secondary ion mass spectrometry* (SIMS) (Section 3.2.5) is in widespread use for the analysis of solid surfaces and materials, including thin films and semiconductors. SIMS is used to detect atomic ions as well as molecular ions, the latter often detected as clusters. Detection limits of the order of 10^{12} and 10^{16} atoms per cubic centimetre. Mass interferences (peaks from different molecules or atoms that share a common *m/z* value) are a common feature of SIMS experiments, and it is necessary to anticipate them in advance such possibilities in the design of an experiment.

9.6.1 Depth Profiling

Since the primary ion beam can be focused to a diameter of less than one μm, SIMS provides a means with which to characterise a surface with high resolution. Where continuous sputtering of the surface is performed, an analysis of the material as a function of depth (a depth profile) can be produced. Typical surface depths are of the order of 1 nm. This analysis is useful in industrial applications in order to study the quality of manufactured coatings or the processes used to construct them. A SIMS depth profile of a stainless steel surface coated with layer of titanium is shown in Figure 9.6. The figure shows that the titanium coating has a depth of 1.4 μm.

One application of SIMS is *ion microscopy*. Here the primary beam is focused on the sample over an area of approximately 10 μm. Secondary ions released from the surface are passed into an electrostatic mirror in which they are energy-focused and reflected back to the mass analyser and onto an image converter. The image converter translates the spatial distribution of the atoms on the surface onto a fluorescent screen for visualisation. In combination with a depth profile, ion microscopy enables three-dimensional maps of a material over a diameter of 250 μm to be constructed with resolutions of the order of 1 μm. Larger surface areas can also be studied though usually at reduced resolution.

9.6.2 Analysis of Impurities

The characterisation of surfaces containing aluminium, silicon, tungsten, gallium and titanium has been accomplished using SIMS including the identification of impurities such as oxygen. The strength

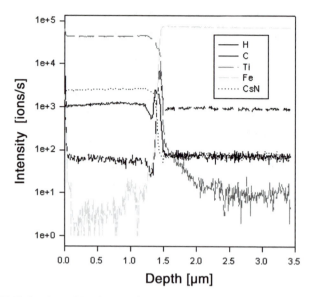

Figure 9.6 *SIMS depth profile of a stainless steel surface coated with layer of titanium.*
The titanium has a depth of 1.4 μm

and mechanical properties of metals and polymers can be substantially
altered by the presence of inert impurities or gas pockets. SIMS has
found particular application in the semiconductor industry since
impurities or doped elements incorporated into the material can
either reduce or enhance the electrical properties of the material.
Semiconductors are usually constructed of single crystals of silicon and
gallium within which undesired impurities must be kept well below the
1% level.

SIMS analysis can determine the distribution of trace levels of
impurities (down to 10 pg per gram of material) in high purity materials
in three-dimensions. In the case of borosilicate glass, it can reveal
impurities such as lithium and sodium at particular depths that can
be responsible for weaknesses and subsequent fractures. It can also be
exploited to examine glass coatings used for optics in scientific and
industrial applications.

9.6.3 Reaction Catalysts

SIMS has been further applied to the study of reaction catalysts in terms
of their molecular structure and that of their clusters. Transition metal
complexes are one type of catalyst that can promote chemical transformations without being consumed during a reaction. The structures

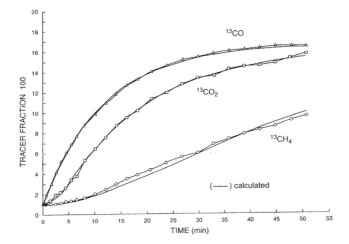

Figure 9.7 *Conversion of carbon monoxide into methane above a nickel catalyst followed by SIMS using a ^{13}C tracer*
(Source: M. Otarod, S. Ozawa, F. Yin, M. Chew, H.Y. Cheh and J. Happel, Multiple isotopic tracing of methanation over nickel catalysts, *J. Catalysis*, 1983, **84**, 156–159)

of metal complexes such as nickel oxide have been studied on polymeric supports and oxide coatings. The poisoning of such catalysts has also been investigated by mass spectrometry.

Nickel oxides act as catalysts to promote the conversion of carbon monoxide and molecular hydrogen into methane above their surface which can be followed by SIMS (Figure 9.7). When hydrogen disulfide is used to "poison" the catalyst, a reduction in the formation of methane can be followed as a function of the proportion of the nickel oxide surface covered with sulphur.

FURTHER READING

J.R. de Laeter, *Applications of Inorganic Mass Spectrometry*, Wiley-Interscience, New York, 2001.

A. Montasser, *Inductively Coupled Plasma Mass Spectrometry*, Wiley-VCH, Berlin, 1998.

J.D. Rosen (ed) *Applications of New Mass Spectrometry Techniques in Pesticide Chemistry*, John Wiley & Sons, New York, 1987.

I.T. Platzner, *Modern Isotope Ratio Mass Spectrometry*, John Wiley & Sons, New York, 1987.

E.R. Badman and R.G. Cooks, Miniature mass analyzers, *J. Mass Spectrom.*, 2000, **35(6)**, 659–671.

T. Kotiaho, F.R. Lauritsen, T.K. Choudhury, R.G. Cooks and G.T. Tsao, Membrane introduction mass spectrometry, *Anal. Chem.*, 1991, **63(18)**, 875A–883A.

R.C. Murphy, G. Preti, M.M. Nafissi-Varchei and K. Biemann, Search for organic material in lunar fines by mass spectrometry, *Science*, 1970, **167(3918)**, 755–7.

K. Biemann, J. Oro, P. Toulmin, L.E. Orgel, A.O. Nier, D.M. Anderson, P.G. Simmonds, D. Flory and A.V. Diaz, Search for organic and volatile inorganic compounds in two surface samples from the Chryse Planitia region of Mars, *Science*, 1976, **194(4260)**, 72–76.

T. Owen, K. Biemann, D.R. Rushneck, J.E. Biller, D.W. Howarth and A.L. LaFleur, The atmosphere of Mars: detection of krypton and xenon, *Science*, 1976, **194(4271)**, 1293–1295.

A.O. Nier, Mass spectrometry in planetary research, *Int. J. Mass Spectrom. Ion Proc.*, 1985, **66**, 55–73.

M.R. Sims, C.T. Pillinger, I.P. Wright, J. Dowson, S. Whitehead, A. Wells, J.E. Spragg, G. Fraser, L. Richter, H. Hamacher, A. Johnstone, N.P. Meredith, C. de la Nougerede, B. Hancock, R. Turner, S. Peskett, A. Brack, J. Hobbs, M. Newns, A. Senior, M. Humphries, H.U. Keller, N. Thomas, J.S. Lingard and T.C. Ng, Beagle 2: a proposed exobiology lander for ESA's 2003 Mars Express mission, *Adv. Space Res.*, 1999, **23(11)**, 1925–1928.

A. Benninghoven, in *Secondary ion mass spectrometry: basic concepts, instrumental aspects, applications, and trends*, A. Benninghoven, F.G.R. Denauer, H.W. Werner (ed), John Wiley & Sons, New York, 1987.

H. Oechsner, Inorganic mass spectrometry for surface and thin film analysis, *Anal. Chim. Acta*, 1993, **283**, 131–138.

J.S. Becker and H.-J. Dietze, Inorganic trace analysis mass spectrometry, *Spectrochimica Acta B*, 1998, **53**, 1475–1506.

J.G. Holland and S.D. Tanner (ed), *Plasma Source Mass Spectrometry: The New Millennium*, Royal Society of Chemistry, Cambridge, UK.

CHAPTER 10

Accelerator Mass Spectrometry

10.1 INTRODUCTION

Accelerator Mass Spectrometry (AMS) is an analytical technique that uses an ion accelerator as an ultrasensitive mass spectrometer to ultimately count individual atoms. AMS was first introduced in 1977 and constitutes a highly sensitive method for detecting very low concentrations of long-lived radioisotopes or stable isotopes in a wide range of samples.

AMS separates rare radioisotopes from stable ones and measures their relative ratio with high sensitivity and precision. It is commonly used in *radiocarbon dating* experiments, where carbon-based materials are converted to graphite, and the amount of ^{14}C they contain is measured. This provides a measure of the age of the item based on the half-life of the ^{14}C isotope of 5568 years. Meteorites from space, air trapped in Antarctic ice, and the Turin shroud are some of the sources of samples to which AMS has been applied. Other applications include studying radiolabelled tracers in biological systems.

In the AMS technique, the element of interest is chemically separated from the original sample and loaded as a target in the sputter ion source of the *tandem accelerator* (Figure 10.1). Samples are pulverised, treated with acid and alkali and freeze-dried. In the case of carbon-based compounds, the sample is converted to either graphite or carbon dioxide. After ionisation of the samples and the separation of ions using a magnet, negative ions containing the radioisotope of interest are accelerated through a potential of several million volts (MV). Negative ions are used to distinguish ^{14}C from ^{14}N since the latter does not form a negative ion. A gas such as sulphur hexafluoride is added to the accelerator to dissociate all molecular ions to an atomic form. At the end of this first acceleration stage these ions pass through a stripper. A stripper consists of a thin carbon foil or gas that strips electrons from ions and destroys any molecular isobars. In the case of carbon, any $^{12}CH_2^+$ and $^{13}CH^+$ ions are fragmented to leave only $^{14}C^+$ ions with a *m/z* of 14. These positive ions are further accelerated to energies of up

175

to several tens of MeV in the second stage of the tandem accelerator. Acceleration of the ions to high energies enables ions to be uniquely identified based on their total energy. Using a magnetic and electrostatic mass analyser, the ions are focused into a Faraday cup for ion detection. For carbon samples, the ratio of $^{14}C/^{13}C$ and $^{14}C/^{12}C$ is measured and compared to measurements made for standards of known ratios. It is possible to measure isotopic ratios down to $1:10^{-15}$, or low attomole levels.

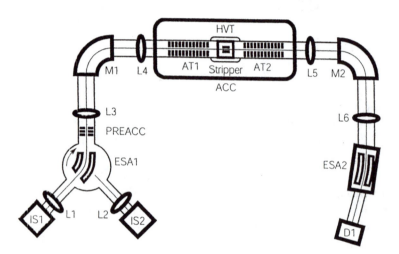

Figure 10.1 *Schematic representation of a tandem accelerator mass spectrometer featuring dual ion sources, preaccelerator and tandem accelerator (ACC)*
(Source: C. Tuniz, J.R. Bird, D. Fink and G.F. Herzog, *Accelerator Mass Spectometry: Ultrasensitive Analysis for Global Science*, CRC Press, Boca Raton, 1998, Ch. 1, p. 42, Fig. 3.1)

AMS is unaffected by almost all background effects that limit conventional mass spectrometry measurements. Thus AMS is five to ten orders of magnitude more sensitive than a conventional mass spectrometry experiment. The amount of sample required for accelerator mass spectrometry is far less (typically a few mg) than that required for beta-particle decay counting, with around 1% of all the ^{14}C in a sample measured.

10.2 ION SOURCES

A typical AMS source consists of a heated reservoir of caesium powder, an ioniser that produces a Cs^+ beam focused at the sample, and an extraction electrode to accelerate and focus secondary negative ions from the sample (Figure 10.2). Solid samples are deposited to a diameter of 1–2 mm in the centre of a cooled metal plate (usually cast of copper or aluminium) which acts as the cathode. The caesium ions are accelerated toward the cathode and on impact sputter or release particles from the surface. Such sputter sources efficiently produce negative ions for many elements and molecules depending on their electron affinity. The negative ions are injected into the accelerator where all molecular ions are dissociated and positive ions are formed.

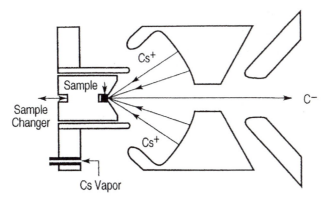

Figure 10.2 *Schematic representation of a typical accelerator mass spectrometry ion source*
(Source: adapted from C. Tuniz, J.R. Bird, D. Fink and G.F. Herzog,
Accelerator Mass Spectometry: Ultrasensitive Analysis for Global Science,
CRC Press, Boca Raton, 1998, Ch. 1, p. 43 – part (b), Fig. 3.2)

10.3 PERFORMANCE AND LIMITATIONS OF RADIOCARBON DATING

Many radioisotopes including ^{3}H, ^{14}C, ^{26}Al, ^{32}Si, ^{39}Ar and ^{81}Kr are produced in the atmosphere by cosmic rays through nuclear reactions. ^{14}C is produced by reactions between stable ^{14}N nuclei and neutrons in the upper atmosphere and is subsequently converted to carbon dioxide. This carbon dioxide is assimilated into plants, and through their consumption into animals. An equilibrium is maintained in the Earth's atmosphere, hydrosphere and biosphere through the continuous production of atmospheric ^{14}C. The stable isotopes of carbon (^{12}C and ^{13}C) constitute the majority of carbon on Earth (98.9% and 1.1% respectively). The level of ^{14}C on Earth, in contrast, is extremely low and has been measured to be about 10^{-10}% of all carbon.

The measurement of ^{14}C is used in numerous applications of which radiocarbon dating is the best known. Radiocarbon dating involves measuring the ^{14}C in biological specimens or archaeological relics to calculate their age. All living organisms or protected archaeological relics contain nearly the same proportion of radioactive carbon at the time of death or burial. The level of ^{14}C subsequently decreases by radioactive decay with a half-life of 5,568 years. By measuring the residual levels of ^{14}C in a sample, the age of the source material can be estimated.

The precision of radiocarbon dating measurements depends on a number of factors including the amounts of material available for analysis, contamination of the sample, reservoir effects, and variations in ^{14}C production. For specimens less than 5000 years of age, a minimum of about 50 µg of sample material is required for analysis. Optimal sample levels are of the order of a few mg to as much as a gram. At the lower levels, errors of the order of 1% in the age in years are typical.

Contamination is a major source of errors in AMS measurements particularly in older specimens. Bone carbonate, for example, is prone to the exchange of carbon from the environment particularly when buried in carbon-rich soils. The ^{14}C content in carbon dioxide trapped in polar ice cores can be measured providing it can be separated from that induced by radiation on the ice surface prior to it becoming buried. The so-called reservoir effect occurs when samples derive carbon not from the Earth's equilibrium but from local environments. These effects are apparent in plants found near volcanoes that release ^{14}C-depleted carbon dioxide and in deep sea aquatic systems.

10.4 APPLICATIONS OF RADIOCARBON DATING IN ARCHAEOLOGY AND COSMOLOGY

The shroud of Turin, an ancient cloth that many Christians believe was used to wrap Christ's body following his crucifixion, represents one of the most controversial radiocarbon dating experiments performed to date. The analysis was performed simultaneously at AMS facilities in Europe and the United States in 1988 from samples cut from the shroud in Turin, Italy. All three laboratories subdivided the samples, and subjected the pieces to different mechanical and chemical cleaning procedures to remove contaminants. The samples were further analysed microscopically to identify and remove any foreign material. All laboratories combusted the textile segments with copper and then converted the resulting CO_2 into graphite targets. Three to five separate measurements were made at each laboratory. The three laboratories in Arizona, Oxford and Zurich reported the age of the shroud at 641 ± 31, 750 ± 30 and 676 ± 24 years respectively, far younger than is possible if the fabric had been used to wrap the body of Jesus Christ (Figure 10.3).

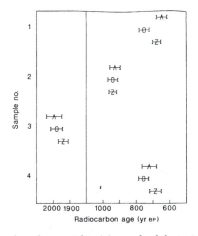

Figure 10.3 *Mean radiocarbon dates, with a ±1 standard deviation of the Shroud of Turin and control samples, as supplied by three laboratories (A, Arizona; O, Oxford; Z, Zurich). The age of the shroud is obtained as AD 1260–1390 with at least 95% confidence*
(Source: P.E. Damon, D.J. Donahue, B.H. Gore, A.L. Hatheway, A.J.T. Jull, T.W. Linick, P.J. Sercel, L.J. Toolin, C.R. Bronk, E.T. Hall, R.E.M. Hedges, R. Housley, I.L. Law, C. Perry, G. Bonani, S. Trumbore, W. Woelfli, J.C. Ambers, S.G.E. Bowman, M.N. Leese and M.S. Tite, Radiocarbon Dating of the Shroud of Turin, *Nature*, 1989, **337(6208)**, 611–615)

Radiocarbon dating measurements can also investigate when early civilisations first occupied land or cultivated crops. Controversy still surrounds the initial occupation of land in Europe, the Americas and Australia. The oldest known human occupation of territories in Australia has been dated at some 40,000 years ago, far before European settlement. Rock art of the indigenous Australian aboriginals uncovered in a series of excavations in far north Queensland has been dated at 26,000 years. Accelerator mass spectrometry has also been used to determine the age of maize cobs in Oaxaco in Mexico. Radiocarbon dating recorded the age of the maize, an ancient corn, at over 6,200 years old, making them the oldest known agricultural crops in the Americas.

The study of extraterrestrial materials from meteorite craters and rock are also of interest to scientists in order to establish that such materials are indeed extraterrestrial. Radiocarbon dating measurements can also be used to determine when such a meteorite collided with Earth and also its exposure to cosmic rays prior to collision in order to suggest its origin. The age of most specimens is evaluated based on ^{14}C and ^{41}Ca levels, the latter with a half-life of 130,000 years. Measurements of

^{41}Ca have the further advantage that its rate of production on Earth varies to a lesser degree than other isotopes.

10.5 BIOMEDICAL APPLICATIONS

^{14}C is used as a radioactive tracer in nuclear medicine both in medical research and for the testing of new pharmaceuticals on volunteers to follow metabolic processes and abnormalities. One method of following the metabolism of a ^{14}C-labelled drug compound, that has been ingested and metabolised, is to collect and analyse the end-product carbon dioxide exhaled. Clinically useful information is usually obtained from carbon dioxide exhaled a few hours after the administration of the drug, even if the time required for its complete metabolism is much longer. This approach has been used to study the long-term retention of ^{14}C-labelled triolein in fat by analysis of the patients' breath. There are, however, considerable errors in the estimates of the absorbed doses of ^{14}C-labelled pharmaceuticals largely due to the long half-life of ^{14}C.

The metabolism of calcium is also of interest in the study of bone diseases such as osteoporosis. An imbalance between the excretion and absorption of calcium from bone is responsible for osteoporosis and can only be partly counteracted by diets rich in calcium due to its absorption through the intestines at approximately 30%. The calcium isotopes ^{45}Ca and ^{47}Ca are usually employed as tracers since their half-lives are far shorter than ^{41}Ca (4.5 and 165 days respectively). In a bone absorption study in menopausal women, volunteers ingested 125 ng of radioactive calcium and their level of uptake was monitored by AMS measurements on their urine. The study's findings revealed a short-term rapid loss of calcium (by three orders of magnitude in 100 days) followed by a period of little loss (900 days).

Studies of the uptake of ^{26}Al from water sources have also been followed in rat brain by AMS in order to quantify the levels of environmental aluminium that enter the blood stream. The rate of passage of ^{26}Al from the blood to the brain is of interest in the study of Alzheimer's disease and other neurological disorders.

Other radioisotopes of importance in biomedical studies include ^{36}Cl and ^{129}I due to the abundance of these elements in insecticides and pesticides. Thus the levels of these compounds absorbed by humans can be measured as a function of exposure.

These studies demonstrate the far-reaching applications of accelerator mass spectrometry. They further illustrate the scope of experimentation and discovery possible a century after the founding of mass spectrometry.

FURTHER READING

C. Tuniz, J.R. Bird, D. Fink and G.F. Herzog, *Accelerator Mass Spectrometry: Ultrasensitive Analysis for Global Science*, Boca Raton, Florida, CRC Press, 1998.

H.E. Gove, *From Hiroshima to the Iceman: The Development and Applications of Accelerator Mass Spectrometry*, Institute of Physics, USA, 1998.

Abbreviations used in Mass Spectrometry

ADC	analog-to-digital converter
AE	appearance energy
AMS	accelerator mass spectrometry
amu	atomic mass unit, now u
AP	appearance potential
APCI	atmospheric-pressure chemical ionisation
B	magnetic sector mass analyser; magnetic field strength (in italics)
B/E	magnetic sector linked scan constant for fragment ions
B^2/E	magnetic sector linked scan constant for precursor or parent ions
CA	collisional activation
CAD	collisionally activated dissociation
CE	capillary electrophoresis
CF-FAB	continuous-flow fast atom (or ion) bombardment
CI	chemical ionisation
CID	collision-induced dissociation
CR	charge reversal
Da	Dalton
DE	delayed extraction, also time-lag focusing (TLF)
E	electric sector mass analyser; electric sector voltage
EA	electron affinity
ECD	electron-capture dissociation
EE	even-electron ion
EI	electron ionisation; electron impact
EM	electron multiplier
ESI	electrospray ionisation

ESI-MS or ESIMS	electrospray ionisation mass spectrum, spectrometry or spectrometer
e	charge of an electron; one electron has a charge of 1.602×10^{-19} Coulombs (C)
eV	electron volt
FA	flowing afterglow
FAB	fast-atom bombardment
FD	field desorption
FFR	field free region
FT-ICR	Fourier-transform ion cyclotron resonance, also FT-MS
FWHM	full-width at half maximum
GC	gas chromatography or chromatograph
GC-MS	gas chromatography-mass spectrometry
ICP-MS	inductively coupled plasma mass spectrometry
ICR	ion cyclotron resonance
IE	ionisation energy
IMS	ion mobility spectrometry
IT	ion trap, also quadrupole ion trap (QIT)
kDa	kilodalton
KER	kinetic energy release
LC-MS	liquid chromatography mass spectrometry or spectrometer
LDI	laser desorption ionisation (also LD)
LSIMS	liquid secondary ion mass spectrometry
MALDI	matrix-assisted laser desorption ionisation
MCI	massive cluster impact
MCP	microchannel plate
MI	metastable ion
MIKES	mass-analysed ion kinetic energy spectrometry
MIMS	membrane introduction mass spectrometry
MRM	multiple-reaction monitoring
MS	mass spectrometer, spectrometry or spectrum
MS/MS	tandem mass spectrometry, spectrometer or spectrum; also MS^2
MW	molecular weight
m/z	mass-to-charge ratio
N	neutral molecule
nanospray	nanolitre flow electrospray ionisation
NR	neutralisation/reionisation
oa	orthogonal acceleration
OE	odd-electron ion
PA	proton affinity

PAD	post acceleration detector
PD	plasma desorption
PID	photon-induced dissociation
PSD	post-source decay
Q	quadrupole mass filter
q	quadrupole mass filter operating in rf-only mode
QET	quasi equilibrium theory
QIT	quadrupole ion trap, also IT
RTOF	reflecting time-of-flight
rf	radio frequency
SELDI	surface-enhanced laser desorption ionisation
SIC	selected ion chromatogram, also extracted ion chromatogram
SID	surface-induced dissociation
SIFT	selected-ion flow tube
SIM	selected-ion monitoring
SIMS	secondary ion mass spectrometry
SRM	selected reaction monitoring
t	ion flight time, time
TIC	total ion current
TLF	time-lag focusing
TOF	time-of-flight
u	unit of mass
V	volt, voltage
z	charge of an ion; an integer multiple of e

Isotope Masses and Abundances

Isotope	Nominal mass	Mass	Relative abundance
^{1}H	1	1.007825032	99.985(1)
^{2}H; D	2	2.014101778	0.015(1)
^{3}H; T	3	3.016049268	<0.0001
^{4}He	4	4.002603250	~100
^{6}Li	6	6.0151223	7.5(2)
^{7}Li	7	7.0160041	92.5(2)
^{9}Be	9	9.0121822	~100
^{10}B	10	10.0129371	19.9(2)
^{11}B	11	11.0093055	80.1(2)
^{12}C	12	12.000000000	98.93(8)
^{13}C	13	13.003354838	1.07(8)
^{14}N	14	14.003074007	99.632(7)
^{15}N	15	15.00010897	0.368(7)
^{16}O	16	15.994914622	99.757(16)
^{17}O	17	16.9991315	0.038(1)
^{18}O	18	17.999160	0.205(14)
^{19}F	19	18.9984032	~100
^{20}Ne	20	19.992440176	90.48(3)

Isotope	Nominal mass	Mass	Relative abundance
^{21}Ne	21	20.99384674	0.27(1)
^{22}Ne	22	21.9913855	9.25(3)
^{23}Na	23	22.9897697	~100
^{24}Mg	24	23.9850419	78.99(4)
^{25}Mg	25	24.9858370	10.00(1)
^{26}Mg	26	25.9825930	11.01(3)
^{27}Al	27	26.9815384	~100
^{28}Si	28	27.97692653	92.22(2)
^{29}Si	29	28.97649472	4.69(1)
^{30}Si	30	29.97377022	3.09(1)
^{31}P	31	30.9737615	~100
^{32}S	32	31.9720707	94.93(31)
^{33}S	33	32.9714585	0.76(2)
^{34}S	34	33.9678668	4.29(28)
^{36}S	36	35.9670809	0.02(1)
^{35}Cl	35	34.96885271	75.78(4)
^{37}Cl	37	36.96590260	24.22(4)
^{36}Ar	36	35.9675463	0.3365(30)
^{38}Ar	38	37.9627322	0.0632(5)
^{40}Ar	40	39.962383123	99.6003(30)
^{39}K	39	38.9637069	93.2581(44)
^{40}K	40	39.9639987	0.0117(1)
^{41}K	41	40.9618260	6.7302(44)
^{40}Ca	40	39.9625912	96.941(156)
^{42}Ca	42	41.9586183	0.647(23)
^{43}Ca	43	42.9587668	0.135(10)

Isotope	Nominal mass	Mass	Relative abundance
^{44}Ca	44	43.955481	2.086(110)
^{46}Ca	46	45.953693	0.004(3)
^{48}Ca	48	47.952533	0.187(21)
^{45}Sc	45	44.955910	~100
^{46}Ti	46	45.952630	8.25(3)
^{47}Ti	47	46.951764	7.44(2)
^{48}Ti	48	47.947947	73.72(3)
^{49}Ti	49	48.947871	5.41(2)
^{50}Ti	50	49.944792	5.18(2)
^{50}V	50	49.947163	0.250(4)
^{51}V	51	50.943964	99.750(4)
^{50}Cr	50	49.946050	4.345(13)
^{52}Cr	52	51.940512	83.789(18)
^{53}Cr	53	52.940653	9.501(17)
^{54}Cr	54	53.938885	2.365(7)
^{55}Mn	55	54.938049	~100
^{54}Fe	54	53.939615	5.845(35)
^{56}Fe	56	55.934942	91.754(36)
^{57}Fe	57	56.935398	2.119(10)
^{58}Fe	58	57.933280	0.282(4)
^{59}Co	59	58.933200	~100
^{58}Ni	58	57.935348	68.0769(89)
^{60}Ni	60	59.930790	26.2231(77)
^{61}Ni	61	60.931060	1.1399(6)
^{62}Ni	62	61.928348	3.6345(17)
^{64}Ni	64	63.927969	0.9256(9)

Isotope	Nominal mass	Mass	Relative abundance
^{63}Cu	63	62.929601	69.17(3)
^{65}Cu	65	64.927794	30.83(3)
^{64}Zn	64	63.929146	48.63(60)
^{66}Zn	66	65.926036	27.90(27)
^{67}Zn	67	66.927131	4.10(13)
^{68}Zn	68	67.924847	18.75(51)
^{70}Zn	70	69.925325	0.62(3)
^{69}Ga	69	68.925581	60.108(9)
^{71}Ga	71	70.924707	39.892(9)
^{70}Ge	70	69.924250	20.84(87)
^{72}Ge	72	71.922076	27.54(34)
^{73}Ge	73	72.923460	7.73(5)
^{74}Ge	74	73.921178	36.28(73)
^{76}Ge	76	75.921403	7.61(38)
^{75}As	75	74.921597	~100
^{74}Se	74	73.922477	0.89(4)
^{76}Se	76	75.919214	9.37(29)
^{77}Se	77	76.919915	7.63(16)
^{78}Se	78	77.917310	23.77(28)
^{80}Se	80	79.916522	49.61(41)
^{82}Se	82	81.916700	8.73(22)
^{79}Br	79	78.918338	50.69(7)
^{81}Br	81	80.916291	49.31(7)
^{78}Kr	78	77.92039	0.35(1)
^{80}Kr	80	79.916379	2.28(6)
^{82}Kr	82	81.913485	11.58(14)

Isotope	Nominal mass	Mass	Relative abundance
^{83}Kr	83	82.914137	11.49(6)
^{84}Kr	84	83.911508	57.00(4)
^{86}Kr	86	85.910615	17.30(22)
^{85}Rb	85	84.911792	72.17(2)
^{87}Rb	87	86.909186	27.83(2)
^{84}Sr	84	83.913426	0.56(1)
^{86}Sr	86	85.909265	9.86(1)
^{87}Sr	87	86.908882	7.00(1)
^{88}Sr	88	87.905617	82.58(1)
^{89}Y	89	88.905849	~100
^{90}Zr	90	89.904702	51.45(40)
^{91}Zr	91	90.905643	11.22(5)
^{92}Zr	92	91.905039	17.15(8)
^{94}Zr	94	93.906314	17.38(28)
^{96}Zr	96	95.908275	2.80(9)
^{93}Nb	93	92.906376	~100
^{92}Mo	92	91.906810	14.84(35)
^{94}Mo	94	93.905087	9.25(12)
^{95}Mo	95	94.905841	15.92(13)
^{96}Mo	96	95.904678	16.68(2)
^{97}Mo	97	96.906020	9.55(8)
^{98}Mo	98	97.905407	24.13(31)
^{100}Mo	100	99.90748	9.63(23)
^{96}Ru	96	95.90760	5.52(20)
^{98}Ru	98	97.90529	1.88(9)
^{99}Ru	99	98.905939	12.74(26)

Isotope	Nominal mass	Mass	Relative abundance
^{100}Ru	100	99.904219	12.60(19)
^{101}Ru	101	100.905582	17.05(7)
^{102}Ru	102	101.904349	31.57(31)
^{104}Ru	104	103.905430	18.66(44)
^{103}Rh	103	102.905504	~100
^{102}Pd	102	101.905607	1.02(1)
^{104}Pd	104	103.904034	11.14(8)
^{105}Pd	105	104.905083	22.33(8)
^{106}Pd	106	105.903484	27.33(3)
^{108}Pd	108	107.903895	26.46(9)
^{110}Pd	110	109.905153	11.72(9)
^{107}Ag	107	106.905093	51.839(8)
^{109}Ag	109	108.904756	48.161(8)
^{106}Cd	106	105.90646	1.25(6)
^{108}Cd	108	107.90418	0.89(3)
^{110}Cd	110	109.903006	12.49(18)
^{111}Cd	111	110.904182	12.80(12)
^{112}Cd	112	111.902758	24.13(21)
^{113}Cd	113	112.904401	12.22(12)
^{114}Cd	114	113.903359	28.73(42)
^{116}Cd	116	115.904756	7.49(18)
^{113}In	113	112.904062	4.29(5)
^{115}In	115	114.903879	95.71(5)
^{112}Sn	112	111.904822	0.97(1)
^{114}Sn	114	113.902783	0.65(1)
^{115}Sn	115	114.903347	0.34(1)

Isotope	Nominal mass	Mass	Relative abundance
116Sn	116	115.901745	14.45(9)
117Sn	117	116.902955	7.68(7)
118Sn	118	117.901608	24.22(9)
119Sn	119	118.903311	8.59(4)
120Sn	120	119.902199	32.59(9)
122Sn	122	121.903441	4.63(3)
124Sn	124	123.905275	5.79(5)
121Sb	121	120.903822	57.21(5)
123Sb	123	122.904216	42.79(5)
120Te	120	119.90403	0.09(1)
122Te	122	121.903056	2.55(12)
123Te	123	122.904271	0.89(3)
124Te	124	123.902819	4.74(14)
125Te	125	124.904424	7.07(15)
126Te	126	125.903305	18.84(25)
128Te	128	127.904462	31.74(8)
130Te	130	129.906223	34.08(62)
127I	127	126.904468	~100
124Xe	124	123.905895	0.09(1)
126Xe	126	125.90427	0.09(1)
128Xe	128	127.903531	1.92(3)
129Xe	129	128.904780	26.44(24)
130Xe	130	129.903509	4.08(2)
131Xe	131	130.905083	21.18(3)
132Xe	132	131.904155	26.89(6)
134Xe	134	133.905395	10.44(10)

Isotope	*Nominal mass*	*Mass*	*Relative abundance*
^{136}Xe	136	135.90722	8.87(16)
^{133}Cs	133	132.905447	~100
^{130}Ba	130	129.90631	0.106(1)
^{132}Ba	132	131.905056	0.101(1)
^{134}Ba	134	133.904504	2.417(18)
^{135}Ba	135	134.905684	6.592(12)
^{136}Ba	136	135.904571	7.854(24)
^{137}Ba	137	136.905822	11.232(24)
^{138}Ba	138	137.905242	71.698(42)
^{138}La	138	137.907107	0.090(1)
^{139}La	139	138.906349	99.910(1)
^{136}Ce	136	135.90714	0.19(1)
^{138}Ce	138	137.90599	0.25(1)
^{140}Ce	140	139.905435	88.48(10)
^{142}Ce	142	141.909241	11.08(10)
^{141}Pr	141	140.907648	~100
^{142}Nd	142	141.907719	27.13(12)
^{143}Nd	143	142.909810	12.18(6)
^{144}Nd	144	143.910083	23.80(12)
^{145}Nd	145	144.912569	8.30(6)
^{146}Nd	146	145.913113	17.19(9)
^{148}Nd	148	147.916889	5.76(3)
^{150}Nd	150	149.920887	5.64(3)
^{144}Sm	144	143.911996	3.1(1)
^{147}Sm	147	146.914894	15.0(2)
^{148}Sm	148	147.914818	11.3(1)

Isotope	Nominal mass	Mass	Relative abundance
^{149}Sm	149	148.917180	13.8(1)
^{150}Sm	150	149.917272	7.4(1)
^{152}Sm	152	151.919729	26.7(2)
^{154}Sm	154	153.922206	22.7(2)
^{151}Eu	151	150.919846	47.8(1.5)
^{153}Eu	153	152.921227	52.2(15)
^{152}Gd	152	151.919789	0.20(1)
^{154}Gd	154	153.920862	2.18(3)
^{155}Gd	155	154.922619	14.80(5)
^{156}Gd	156	155.922120	20.47(4)
^{157}Gd	157	156.923957	15.65(3)
^{158}Gd	158	157.924101	24.84(12)
^{160}Gd	160	159.927051	21.86(4)
^{159}Td	159	158.925343	~100
^{156}Dy	156	155.92428	0.06(1)
^{158}Dy	158	157.924405	0.10(1)
^{160}Dy	160	159.925194	2.34(6)
^{161}Dy	161	160.926930	18.9(2)
^{162}Dy	162	161.926795	25.5(2)
^{163}Dy	163	162.928728	24.9(2)
^{164}Dy	164	163.929171	28.2(2)
^{165}Ho	165	164.930319	~100
^{162}Er	162	161.928775	0.14(1)
^{164}Er	164	163.929197	1.61(2)
^{166}Er	166	165.930290	33.6(2)
^{167}Er	167	166.932046	22.95(15)

Isotope	*Nominal mass*	*Mass*	*Relative abundance*
^{168}Er	168	167.932368	26.8(2)
^{170}Er	170	169.935461	14.9(2)
^{169}Tm	169	168.934211	~100
^{168}Yb	168	167.933895	0.13(1)
^{170}Yb	170	169.934759	3.05(6)
^{171}Yb	171	170.936323	14.3(2)
^{172}Yb	172	171.936378	21.9(3)
^{173}Yb	173	172.938207	16.12(21)
^{174}Yb	174	173.938858	31.8(4)
^{176}Yb	176	175.942569	12.7(2)
^{175}Lu	175	174.940768	97.41(2)
^{176}Lu	176	175.942683	2.59(2)
^{174}Hf	174	173.940042	0.162(3)
^{176}Hf	176	175.941403	5.206(5)
^{177}Hf	177	176.943220	18.606(4)
^{178}Hf	178	177.943698	27.297(4)
^{179}Hf	179	178.945815	13.629(6)
^{180}Hf	180	179.946549	35.100(7)
^{180}Ta	180	179.947466	0.012(2)
^{181}Ta	181	180.947996	99.988(2)
^{180}W	180	179.946706	0.120(1)
^{182}W	182	181.948205	26.498(29)
^{183}W	183	182.950224	14.314(4)
^{184}W	184	183.950932	30.642(8)
^{186}W	186	185.954362	28.426(37)
^{185}Re	185	184.952955	37.40(2)

Isotope	Nominal mass	Mass	Relative abundance
^{187}Re	187	186.955751	62.60(2)
^{184}Os	184	183.952491	0.020(3)
^{186}Os	186	185.953838	1.58(10)
^{187}Os	187	186.955748	1.6(1)
^{188}Os	188	187.955836	13.3(2)
^{189}Os	189	188.958145	16.1(3)
^{190}Os	190	189.958445	26.4(4)
^{192}Os	192	191.961479	41.0(3)
^{191}Ir	191	190.960591	37.3(5)
^{193}Ir	193	192.962923	62.7(5)
^{190}Pt	190	189.95993	0.01(1)
^{192}Pt	192	191.961035	0.79(6)
^{194}Pt	194	193.962663	32.9(6)
^{195}Pt	195	194.964774	33.8(6)
^{196}Pt	196	195.964934	25.3(6)
^{198}Pt	198	197.967875	7.2(2)
^{197}Au	197	196.966551	~100
^{196}Hg	196	195.965814	0.15(1)
^{198}Hg	198	197.966752	9.97(8)
^{199}Hg	199	198.968262	16.87(10)
^{200}Hg	200	199.968309	23.10(16)
^{201}Hg	201	200.970285	13.18(8)
^{202}Hg	202	201.970625	29.86(20)
^{204}Hg	204	203.973475	6.87(4)
^{203}Tl	203	202.972329	29.524(14)
^{205}Tl	205	204.974412	70.476(14)

Isotope	Nominal mass	Mass	Relative abundance
^{204}Pb	204	203.973028	1.4(1)
^{206}Pb	206	205.974449	24.1(1)
^{207}Pb	207	206.975880	22.1(1)
^{208}Pb	208	207.976636	52.4(1)
^{209}Bi	209	208.980384	~100
^{232}Th	232	232.038050	~100
^{234}U	234	234.040945	0.0055(5)
^{235}U	235	235.043922	0.720(1)
^{238}U	238	238.050784	99.2745(15)

(Data source: Table of the Isotopes (Revised 1998), Norman E. Holden, Brookhaven National Laboratory, New York U.S.A. Data edited and compiled by Jason W.H. Wong, University of Sydney)

Comparison of Common Ionisation Techniques*

Ionisation technique	Symbol	Mass limit	Types of molecules	Advantages
Electron ionisation	**EI**	500	Volatile organics, gases, non-polar compounds	Provides structural information as well as molecular weight
Chemical ionisation	**CI**	500	Volatile organics, gases, non-polar compounds	Easy to implement on EI source, enhances molecular ion production
Fast atom bombardment	**FAB**	20,000	Polar compounds including small biopolymers	Modest ionisation efficiencies for large molecular weight compounds
Matrix-assisted laser desorption ionisation	**MALDI**	up to 1,000,000	Polar compounds from small molecules to large biopolymers	Easy to perform, suited to high throughput, high ionisation efficiencies
Electrospray ionisation	**ESI**	up to 5,000,000	Polar compounds from small molecules to large biopolymers	High ionisation efficiencies, compatible with LC and CE separation

* Descriptions are to be used as a guide only.

Comparison of the Performance of Mass Analysers*

Analyser	Symbol	Dynamic mass range	Mass resolution	Advantages
Time-of-flight	**TOF**	unlimited	500 (linear) 1000 (reflector) 5,000 (with time-lag focusing)	Relatively easy to construct, Inexpensive, non-scanning, high ion transmission
Magnetic sector	**B**	10,000	10,000	High resolution, high mass accuracy
Quadrupole	**Q**	up to 5000	2000	Inexpensive, tolerant of higher pressures, fast-scanning capability
Quadrupole ion trap	**QIT**	up to 5000	2000	Inexpensive, compact, fast-scanning capability
Ion cyclotron resonance	**ICR**	10,000	up to 500,000	Very high resolution and mass accuracy

*Values are to be used as a guide only. Values vary among instruments. Mass resolutions quoted represent 10% valley definitions

APPENDIX 5

Common Neutral Losses during the Fragmentation of Organic Compounds*

Compound class	Nominal mass loss	Formula
Aliphatic hydrocarbons	15, 29, 43, 57, etc. 28	$-CH_3^{\bullet}$, $-CH_3CH_2^{\bullet}$, $-CH_3(CH_2)_n^{\bullet}$, etc. $-CH_2=CH_2$
Aliphatic alcohols	1, 2, 15, 29, 43 etc. 18 28	$-H \bullet$, $-H_2$, $-CH_3^{\bullet}$, $-CH_3CH_2^{\bullet}$, etc. $-H_2O$ $-CH_2=CH_2$
Phenols	as above, and 28, 29	$-CO$, $-CHO^{\bullet}$
Ethers	31, 45, 59, etc.	$-CH_3O^{\bullet}$, $-CH_3CH_2O^{\bullet}$, $-CH_3(CH_2)_nO^{\bullet}$, etc.
Aliphatic amines	17 15, 29, 43, 57, etc.	$-NH_3$ $-CH_3^{\bullet}$, $-CH_3CH_2^{\bullet}$, $-CH_3(CH_2)_n^{\bullet}$, etc.
Aldehydes and ketones	1, 15, 29, 43, 57, etc.	$-H^{\bullet}$, $-CH_3^{\bullet}$, $-CH_3CH_2^{\bullet}$, $-CH_3(CH_2)_n^{\bullet}$, etc. (adjacent to C=O group)
Carboxylic acids, esters and amides	28, 43, 57, etc. 16 or 17 or 31, 45, 59, etc.	$-CH_2=CH_2$, $-R-CH_2=CH_2$, etc. via a McLafferty rearrangement $-OH$ (acid), $-OR$ (ester) or $-NH_2$ (amide)
Alkyl halides	19, 35/37, 79/81, 129 20, 36/38 15, 29, 43, 57, etc.	$-F^{\bullet}$, $-Cl^{\bullet}$, $-Br^{\bullet}$, $-I^{\bullet}$ (favoured for Br and I) $-HF$, $-HCl$ $-CH_3^{\bullet}$, $-CH_3CH_2^{\bullet}$, $-CH_3(CH_2)_n^{\bullet}$, etc. from cleavage α to the halide atom

*Fragmentation processes to be used as a guide for the interpretation of EI mass spectra. Refer to Chapter 5 for more detail

APPENDIX 6

Summary of Common Fragment Ions Detected for Organic Compounds by Class*

Compound class	Representative formula	m/z of fragment ions
Aliphatic hydrocarbons	C_nH_{2n+2}	15, 29, 43, 57, 71, 85, 99, 113
Aliphatic alcohols or ethers	$C_nH_{2n+1}OH/R$	31, 45, 59, 73, 87, 101
Aliphatic amines	$C_nH_{2n+1}NH_2$	30, 44, 58, 72, 86, 100, 114
Aldehydes and ketones	$C_nH_{2n+1}COH/R$	29, 43, 57, 71, 85, 99
Carboxylic acids and esters	$C_nH_{2n-1}O_2H/R$	45 (acid only), 59, 73, 87, 101
Amides	$C_nH_{2n+1}CONH_2$	44, 58, 72, 86, 100
Alkyl halides	$C_nH_{2n+1}X$	33(F), 49/51(Cl), 93/95(Br), 143(I) 42, 56, 70, 84, 98, 112

*Fragment ions are to be used as a guide for the interpretation of EI mass spectra. Refer to Chapter 5 for more detail.

Gas Phase Acidity Data

Acid M$\underline{H}$	ΔH_{acid}^{0} (kJ mol)
MeC≡CH	1 589.31
MeOH	1 587.63
EtOH	1 574.66
HC≡CH	1 571.72
iPrOH	1 566.28
iBuOH	1 563.35
tBuOH	1 562.93
tBuCH$_2$O$\underline{H}$	1 556.65
HF	1 555.40
PhCH$_2$O$\underline{H}$	1 547.44
nPrOH	1 541.16
PhNH$_2$	1 536.97
CH$_3$SO$_2$CH$_3$	1 534.88
CH$_3$C$\underline{H}$O	1 534.04
CF$_3$C$\underline{H}_2$O$\underline{H}$	1 525.67
pyrrole	1 510.18
MeSH	1 503.06
EtSH	1 496.36
nPrSH	1 492.17
tBuSH	1 485.06
H$_2$S	1 479.62
HCN	1 478.36
pMeOC$_6$H$_4$O$\underline{H}$	1 472.50
CF$_3$C(=O)CH$_3$	1 466.64
PhOH	1 464.54
MeCO$_2$H	1 459.10
EtCO$_2$H	1 454.08
PhCO$_2$H	1 418.49
HCl	1 408.86
CF$_3$CO$_2$H	1 351.08

(Data sourced from J.E. Bartmess and R.T. McIver Jr., The Gas Phase Acidity Scale, in *Gas Phase Ion Chemistry*, Vol. 2, Academic Press, New York, 1979. Data is derived from pulsed ICR measurements at 298 K and high pressure MS measurements (shaded) at 500–600 K)

Amino Acid Residue Masses and Modifying Groups

Residue	Code	Elemental compostion	Monoisotopic mass	Average mass
Alanine	A	C_3H_5NO	71.03711	71.0788
Cysteine	C	C_3H_5NOS	103.00919	103.1448
Aspartic acid	D	$C_4H_5NO_3$	115.02694	115.0886
Glutamic acid	E	$C_5H_7NO_3$	129.04259	129.1155
Phenylalanine	F	C_9H_9NO	147.06841	147.1766
Glycine	G	C_2H_3NO	57.02146	57.0520
Histidine	H	$C_6H_7N_3O$	137.05891	137.1412
Isoleucine	I	$C_6H_{11}NO$	113.08406	113.1595
Lysine	K	$C_6H_{12}N_2O$	128.09496	128.1742
Leucine	L	$C_6H_{11}NO$	113.08406	113.1595
Methionine	M	C_5H_9NOS	131.04049	131.1986
Asparagine	N	$C_4H_6N_2O_2$	114.04293	114.1039
Proline	P	C_5H_7NO	97.05276	97.1167
Glutamine	Q	$C_5H_8N_2O_2$	128.05858	128.1742
Arginine	R	$C_6H_{12}N_4O$	156.10111	156.1876
Serine	S	$C_3H_5NO_2$	87.03203	87.0782
Threonine	T	$C_4H_7NO_2$	101.04768	101.1051
Valine	V	C_5H_9NO	99.06841	99.1326
Tryptophan	W	$C_{11}H_{10}N_2O$	186.07931	186.2133
Tyrosine	Y	$C_9H_9NO_2$	163.06333	163.1760

Modifying group	Elemental composition	Monoisotopic mass	Average mass
Hydrogen	H	1.00782	1.0079
Methyl	CH_3	15.02347	15.0348
Formyl	CHO	29.00274	29.0183
Acetyl	C_2H_3O	43.01839	43.0452
t-Butyl	C_4H_9	57.07042	57.1154
Hydroxyl	HO	17.00274	17.0073
Amide	NH_2	16.01872	16.0226
Oxidation	O	15.99491	15.9994
Phosphorylation	PO_3	79.96633	79.9799
Sulphation	SO_3	79.95682	80.0642
Carboxyamidomethyl	C_2H_4NO	57.02146	57.0520

Mononucleotide Residue Masses

Nucleotide	Code	Elemental composition	Monoisotopic mass	Average mass
Adenosine	**A**	$C_{10}H_{12}N_5O_6P$	329.05252	329.2091
Guanosine	**G**	$C_{10}H_{12}N_5O_7P$	345.04744	345.2085
Cytosine	**C**	$C_9H_{12}N_3O_7P$	305.04129	305.1841
Uracil	**U**	$C_9H_{11}N_2O_3P$	306.02530	306.1688
Deoxyadenosine	**dA**	$C_{10}H_{13}N_2O_7P$	313.05761	313.2097
Deoxyguanosine	**dG**	$C_{10}H_{12}N_3O_5P$	329.05252	329.2091
Deoxycytidine	**dC**	$C_9H_{12}N_3O_6P$	289.04637	289.1847
Thymidine	**T**	$C_{10}H_{13}N_2O_7P$	304.04604	304.1963

Monosaccharide Residue Masses

Residue	Abbreviation	Monoisotopic mass	Average mass
Pentoses	**Ara** **Rib** **Xyl**	132.04226	132.1161
Deoxyhexoses	**Fuc** **Rha**	146.05791	146.1430
Hexosamines	**GalN** **GlcN**	161.06881	161.1577
Hexoses	**Fru** **Gal** **Glc** **Man**	162.05282	162.1424
Glucuronic acid	**HexA**	176.0321	176.1259
N-Acetylhexosamines	**GalNAc** **GlcNAc**	203.07937	203.1950
N-Acetylneuraminic acid	**NeuAc**	291.09542	291.2579
N-Glycolylneuraminic acid	**NeuGc**	307.09033	307.2573

Web Sites on Mass Spectrometry

Resources

i-mass.com   *http://www.i-mass.com*
Ion Source   *http://www.ionsource.com*
Spectroscopy Now (incorporating Base Peak)   *http://www.spectroscopynow.com*

Societies

American Society for Mass Spectrometry   *http://www.asms.org*
Australian & New Zealand Society for Mass Spectrometry
                                        http://www.latrobe.edu.au/anzsms/
British Mass Spectrometry Society   *http://www.bmss.org.uk*
International Mass Spectrometry Society   *http://www.imss.nl*
Mass Spectrometry Society of Japan   *http://www.mssj.jp*

Chemical and Biological Online Tools

Isotope Pattern Calculator   *http://www.shef.ac.uk/~chem/chemputer/isotopes.html*
FindMod tool   *http://au.expasy.org/tools/findmod/*
GlycoSuite at Proteome Systems   *https://tmat.proteomesystems.com/glycosuite/*
Mascot Search – Matrix Science   *http://www.matrixscience.com/search_form_select.html*
Peptide Search at EMBL
        http://www.mann.embl-heidelberg.de/GroupPages/PageLink/peptidesearchpage.html
Protein Prospector at UCSF   *http://prospector.ucsf.edu/*
PROWL at Rockefeller   *http://prowl.rockefeller.edu/*
Web Elements Periodic Table   *http://www.webelements.com/*

Tutorials

ASMS – What is Mass Spectrometry?   *http://www.asms.org/whatisms/*
BMSS – What is Mass Spectrometry?
                        http://www.bmss.org.uk/what_is/whatisframeset.html
Cambridge University tutorial   *http://www-methods.ch.cam.ac.uk/meth/ms/theory/*
i-mass guides   *http://www.i-mass.com/guide/*

Subject Index